The Michigan Historical Reprint Series

Reprints from the collection of the University of Michigan University Library

This volume is produced from digital images created through the University of Michigan University Library's preservation reformatting program. The Library seeks to preserve the intellectual content of items in a manner that facilitates and promotes a variety of uses. The digital reformatting process results in an electronic version of the text that can both be accessed online and used to create new print copies. This book and thousands of others can be found in the digital collections of the University of Michigan Library. The University Library also understands and values the utility of print, and makes reprints available through its Scholarly Publishing Office.

For access to the University of Michigan Library's digital collections, please see http://www.lib.umich.edu.

The Scholarly Publishing Office seeks to disseminate high-quality, cost-effective scholarly content through both print and electronic publishing. Information about the Scholarly Publishing Office can be found at http://spo.umdl.umich.edu.

The Scholarly Publishing Office
the University of Michigan
University Library

THE INTERNATIONAL SCIENTIFIC SERIES

VOLUME XIII.

THE INTERNATIONAL SCIENTIFIC SERIES.

Works already Published.

I. THE FORMS OF WATER IN RAIN AND RIVERS, ICE AND GLACIERS. By J. TYNDALL, LL. D., F. R. S. With 26 Illustrations. Price, $1.50.

II. PHYSICS AND POLITICS; OR, THOUGHTS ON THE APPLICATION OF THE PRINCIPLES OF "NATURAL SELECTION" AND "INHERITANCE" TO POLITICAL SOCIETY. By WALTER BAGEHOT. Price, $1.50.

III. FOODS. By DR. EDWARD SMITH. Illustrated. Price, $1.75.

IV. MIND AND BODY: THE THEORIES OF THEIR RELATIONS. By ALEXANDER BAIN, LL. D. Price, $1.50.

V. THE STUDY OF SOCIOLOGY. By HERBERT SPENCER. Price, $1.50.

VI. THE NEW CHEMISTRY. By PROFESSOR JOSIAH P. COOKE, of Harvard University. Illustrated. Price, $2.00.

VII. ON THE CONSERVATION OF ENERGY. By PROFESSOR BALFOUR STEWART. Fourteen Engravings. Price, $1.50.

VIII. ANIMAL LOCOMOTION; OR, WALKING, SWIMMING, AND FLYING. By DR. J. B. PETTIGREW, M. D., F. R. S. 119 Illustrations. Price, $1.75.

IX. RESPONSIBILITY IN MENTAL DISEASE. By DR. HENRY MAUDSLEY. Price, $1.50.

X. THE SCIENCE OF LAW. By PROFESSOR SHELDON AMOS. Price, $1.75.

XI. ANIMAL MECHANISM; OR, AËRIAL AND TERRESTRIAL LOCOMOTION. By C. J. MAREY, Professor of the College of France; Member of the Academy of Medicine, Paris. 117 Engravings. Price, $1.75.

XII. HISTORY OF THE CONFLICT BETWEEN RELIGION AND SCIENCE. By JOHN W. DRAPER, M. D., LL. D. Price, $1.75.

THE INTERNATIONAL SCIENTIFIC SERIES.

THE DOCTRINE OF DESCENT AND DARWINISM.

BY

OSCAR SCHMIDT,

PROFESSOR IN THE UNIVERSITY OF STRASBURG.

WITH TWENTY-SIX WOODCUTS.

NEW YORK:
D. APPLETON AND COMPANY,
549 & 551 BROADWAY.
1875.

PREFACE.

THE important chapter which closes this work was included in a public lecture which I delivered at the meeting of Naturalists and Physicians held this year at Wiesbaden; my purpose being, as I am willing to confess, to ascertain, by experience, whether on this significant subject I had struck the right note to suit a circle of hearers and readers not hampered by prejudice.

After the reception given to this fragment, which I also issued in a separate pamphlet, under the title of "The Doctrine of Descent, in its application to Man" (Die Anwendung der Descendenzlehre auf den Menschen), I venture to hope that the whole may find a welcome.

With the exception of the Ecclesiastico-political question, no sphere of thought agitates the educated classes of our day so profoundly as the doctrine of descent. On both subjects the cry is, "Avow your colours!"

We have, therefore, endeavoured to define our standpoint sharply in the introduction, and to preserve it rigidly throughout the work. This is, indeed, a case in which, as Theodor Fechner has recently said, a definite decision has to be made between two fundamental alternatives. May our exposition afford a lucid testimony to this dictum of one of the patriarchs of the philosophical view of nature.

STRASBURG, *October 18th*, 1873.

OSCAR SCHMIDT.

CONTENTS.

I.

PAGE

INTRODUCTION—SUMMARY OF THE RESULTS OF LINGUISTIC INQUIRY —POSITIVE KNOWLEDGE PRELIMINARY TO THE DOCTRINE OF DESCENT—BELIEF IN MIRACLE—THE LIMITS OF THE INVESTIGATION OF NATURE 1

II.

THE ANIMAL WORLD IN ITS PRESENT STATE 24

III.

THE PHENOMENA OF REPRODUCTION IN THE ANIMAL WORLD. . 39

IV

THE ANIMAL WORLD IN ITS HISTORICAL AND PALÆONTOLOGICAL DEVELOPMENT 60

V.

THE STANDPOINT OF THE MIRACULOUS, AND THE INVESTIGATION OF NATURE—CREATION OR NATURAL DEVELOPMENT—LINNÆUS —CUVIER—AGASSIZ—EXAMINATION OF THE IDEA OF SPECIES . 82

VI.

NATURAL PHILOSOPHY—GOETHE—PREDESTINED TRANSFORMATION ACCORDING TO RICHARD OWEN—LAMARCK 104

VII.

PAGE

Lyell and Modern Geology—Darwin's Theory of Selection —Beginning of Life 127

VIII.

Heredity—Reversion—Variability—Adaptation—Results of Use and Disuse of Organs—Differentiation leading to Perfection 165

IX.

The Development of the Individual (Ontogenesis) is a Repetition of the Historical Development of the Family (Phylogenesis) 195

X.

The Geographical Distribution of Animals in the light of the Doctrine of Derivation 222

XI.

The Pedigree of Vertebrate Animals 248

XII.

Man 282

References and Quotations 311

LIST OF ILLUSTRATIONS.

FIG.		PAGE
1.	Legs of Bird and Chick	9
2.	Medusa, Tiaropsis Diadema	31
3.	Stauridium. Medusa, Cladonema Radiatum	43
4.	Spermatozoa	45
5.	Section of Larva of Calcareous Sponge	51
6.	Embryo of Hydrophilus Piceus	53
7.	Sessile Stage of Crinoid	56
8.	Larva of Crayfish	56
9.	Graptolites	69
10.	Trilobites remipes	70
11.	Palæoniscus	72
12.	Larva of Echinoderm	197
13.	Larva of Sea-Snail	200
14.	Stauridium. Cladonema	203
15.	Hydractinea carnea	204
16.	Larva of Parasitic Crustacea	207
17.	Axolotl	208
18.	Amblystoma	209
19.	Ammonites Humphresianus	214
20.	Ancyloceras	216
21.	Section of Larva of Calcareous Sponge	217
22.	Larva of Lancelet	251
23.	Larva of Ascidian	253
24.	Full-grown Ascidian	255
25.	Impression of Tail of Archæopteryx	266
26.	Skeletons of Feet: Anchitherium, Hipparion, Horse	274

THE DOCTRINE OF DESCENT AND DARWINISM.

I.

Introduction—Summary of the Results of Linguistic Inquiry—Positive Knowledge preliminary to the Doctrine of Descent—Belief in Miracle—The Limits of the Investigation of Nature.

A CRAVING to understand existence pervades mankind, and the life of every self-conscious individual. Every system of philosophy has endeavoured to penetrate into the nature of things, and has originated in the attempt to apprehend the coherency of those great series of material and spiritual phenomena, of which man flatters himself that he is the centre or the end.

Some quiet themselves by emphasizing the contrast between mind and body, idea and phenomenon; others, by the catchword of identity; some have deemed themselves and the world in the most beautiful harmony; others, from the times of the Buddhists, in the 6th century B.C., to the eccentric saints of the present day, the followers and reformers of Schopenhauer's system, regard the world as a mere accumulation of discomfort and conflict, from which the sage may escape by a complete withdrawal into himself, and a return, by the force of an iron will, to an absence of needs and to nothingness.

In all these endeavours to be reconciled and contented with the world, the consciousness of man has made no very important progress. Marvellous as are the attainments of our generation, whether in the domain of individual sciences, or in the sphere of commerce and industry, it is scarcely less wonderful how little certain or advanced is the opinion of the multitude on general questions. Even now, as much as in the days of Aristophanes, the multitude, and likewise many men of "culture," allow themselves to be imposed upon by empty jargon. We no longer burn witches, but verdicts of heresy still abound. As the basis of scientific medicine, our experimental physiology enjoys unexampled encouragement, and a general instinctive recognition unparalleled in former times; but these do not prevent the door from remaining open, in all classes of society, to the most audacious quackery.

We have only to look round at the spiritualists and summoners of souls, who now form special sects and societies; at the advocates of cures by sympathy and incantation, and we can but marvel at the extensive sway of a superstition hardly superior to the Fetichism of a race so alien to ourselves as are the negroes. These are only individual cases of the very widespread lack of judgment, which prevails wherever the supposed enigma of human existence is concerned. Millions and millions who would turn away indignantly if required to believe that anything not entirely natural occurred in the most complicated machine, in the most elaborate product of the chemical retort, or in the strangest results of physical experiment, are yet disposed to seek a dualism behind the processes of life. Wherever, also, the explanation of life, and the reduction of vital phenomena to their true natural causes is concerned, they would wish to deny point-blank the possibility of such explanation or such knowledge, and to refer life to an unapproachable and mystic domain. Or, if the solution of the problem of life be admitted in the abstract, at least something peculiar, and a different standard from that by which other living beings may be measured, is required for the beloved Self.

If we thus see, on the one side, a great portion of our contemporaries either standing before the most important of all problems in utter perplexity and helplessness, or solving it by the theology of revelation, we may, fortunately, point, on the other side, to the goodly host of those who, since the development of science has admitted of it, have encountered the investigation of man's place in nature with sincere interest, and have weighed the problem with intelligence.

This craving for a knowledge based on philosophical and natural science, became apparent about a century ago, and coincided with the first beginnings of linguistic science. It is the more appropriate to allude here to this, as the theories of the origin of language are profoundly affected and influenced by opinions as to the origin of Man, and *vicè versâ.*

The result of an inquiry, made in 1580, as to the language of Paradise, having been that God spoke Danish, Adam Swedish, and the serpent French, Leibnitz, in his letters to Newton, first attempted to regulate the method of linguistic research by recommending as its basis the study of the more recent and known languages. And when, in the middle of last century, the two opinions, that language was invented or revealed, were sharply opposed to each other, and when Süssmilch (1764), in contradiction to Maupertuis and Jean J. Rousseau, had established that invention was not possible without thought, nor thought without language, and, therefore, that the invention of language was a self-contradiction, Herder opportunely entered the lists with his work on language (1770), which formed an epoch in the science.

According to him, language begins with imitations of sounds, at first almost unconscious; the tokens, as he expresses it, by which the soul distinctly recalls an idea. He makes language develop itself from the crudest beginnings, by the increasing need of such verbal tokens; and shows that with the development of mankind, the store of words must also have unconsciously and instinctively increased. The multiplicity of languages is due to the dispersion of nations, whose idiosyncrasies are reflected in the various languages. Thus Herder long ago pointed out the importance of a psychology of nations. He was joined by Wilhelm von Humboldt, whose opinions form the basis of the present science of language, and who held that the imitations of sounds are instinctively crystallized into words, and that with this formation of words and language thought commences. It follows from the nature of these beginnings, that language is the natural expression of the spirit of a people; that it does not stand still, but is for ever in process of transformation.

The science of language, with its great results, displays the most important side of human nature—man in the elevation which he has gradually acquired above the rest of the living world—but it displays this side alone. Although the founders of linguistic inquiry, of whom we have already spoken, had already represented man as first acquiring reason and becoming man, by means of language proceeding from primitive rudiments, they were, nevertheless, satisfied to assume the privileged position of man as an absolute endowment, or a self-evident axiom. This continued as long as natural science was limited to a merely superficial classification of organisms.

Man, as consisting of flesh and blood, seemed, indeed, akin to the higher animals; but so long as their descent, their actual consanguinity was not discussed, so long as nothing was demanded beyond their juxtaposition, according to the analogy of their characteristics, without any scrutiny of the deeper causes of their divergence or similarity, man indisputably occupied the highest grade in the system of living beings. Linnæus places man in the order of Primates, together with bats, lemurs, and apes, without, on that account, being accused from pulpit and from chair of an assault on the dignity of mankind. Buffon, likewise, was able, unrebuked, to indulge his whim, by specially discussing our race in his description of the ass.

Only when, quite recently, the world became aware that the word "affinity," hitherto uttered with supreme indifference, was henceforth to be taken seriously and literally, since that which is akin is also the fruit of one and the same tree, a beam of joyful recognition thrilled through those to whom man appeared a being completely within the bounds of nature. But others, who can think of man only as a being absolutely endowed above his natural surroundings, could not fail to regard as a sort of crime the deduction which an all-embracing theory applied with relentless logic to man.

The interest with which the modern theory of kindred and descent has been received does not, therefore, proceed from friends alone, but quite as much from antagonists, who perceive, more or less distinctly, the danger with which the new doctrine threatens their standpoint of miracle.

Even in England the opposition to the great Englishman, with whose name the revolution is connected, has been very considerable, especially since it became evident that, true to himself, he includes man also within the range of his researches, and purposes to apply to him all the consequences of his doctrine. But it appears to me that the dispute and the agitation are still keener on this side of the channel, where Darwinism is meat and drink to the daily papers, and to the philosophical and theological periodicals.

This phenomenon is obvious to all eyes, and we are convinced of the deep importance of the subject which, whether we take part for, or against it, must influence our whole theory of life. Here too that has happened to many, which so often happens in questions the difficulties of which are veiled by an apparent general familiarity. Every one thinks himself capable of deciding about life, and, since to non-scientific persons the notorious relationship with apes is the alpha and omega of the doctrine of Descent—since the most confused heads are often most thoroughly convinced of their own pre-eminence—on no subject do we so frequently hear superficial opinions, mostly condemnatory, and all evincing the grossest ignorance.

I wish then to render the reader able to survey the whole ramified and complicated problem of the doctrine of Descent, and its foundation by Darwin, and to enable him to understand its cardinal points. But we must first dispose of a preliminary question of universal importance and special significance, which is frequently ignored by philosophical and theological opponents, that is, the question of the limits of the investigation of nature. For if it were an established principle that the mystery of the living is different from that of the non-living, that the former might be disclosed, but that the latter is shrouded in a veil which never can be raised, as is even now so frequently asserted, then, indeed, all research directed towards the comprehension of life would be utterly vain and hopeless.

But if the possibility of investigating life and its origin be not opposed by any *à priori* scruples, still more, if the limits of investigation and knowledge, which undoubtedly exist, are no other for animate nature than for the inanimate world of matter, we may venture to approach our task. This will be most adequately effected by making ourselves somewhat familiar with the object of the doctrine of Descent, restricting ourselves, however, to the animal world. If I say then that we must obtain a foundation for the theory of derivation or descent, for the doctrine of the gradual and direct development of the higher and now-existing organisms from lower ancestral forms—in short, for the doctrine of the continuity of life, we must begin with a survey of the animal forms now spread over the earth. As astronomy begins with the mere classification of the stars and constellations, and the knowledge of their apparent motions, so do we also range our material in large groups, and this in the manner offered by the historical development of science.

What first strikes the observer of the animal world is, that it consists of apparently innumerable forms. The primary requirement is discrimination and arrangement. In the first stages of their development, zoology, as well as botany and mineralogy, necessarily consisted of mere descriptions, of a knowledge of objects in a state of completeness. Physics and chemistry, on the other hand, deal with the investigation of phenomena directly referring to their origin, that is to say, with series of phenomena mutually connected as causes and effects, the knowledge of which, therefore, leads at once to results satisfactory and tranquillizing to the mind. This description, at first limited to the exterior, was gradually extended to the interior, because zootomy and comparative anatomy, even more than fifty years ago, had advanced so far in the accumulation of endless details that Cuvier then ventured to found the Natural System.

But this delineation of the animal world required completion on two sides, and, as the science proceeded towards perfection, it received it almost simultaneously on both. To the knowledge of the existence of an animal belongs also the description of its origin. I say emphatically, "the description," for the history of animal development is not as yet in itself a natural science in the same sense as the mathematico-physical sciences; it is a mere description of nature. But it yields a far more accurate knowledge. In many cases it discloses, for the first time, the significance of organs, and gives to comparative anatomy the confirmation, and frequently the possibility of interpretation. The wing of a bird, in its individual parts, may be traced back without difficulty to the anterior extremities of a reptile or a mammal. But the leg of a bird, as a complete organ does not harmonize with the leg of other vertebrata until the development of the bird in the egg reveals that the disposition of the segments and of the articulations is precisely the same in both cases, and that the apparent anomaly is produced merely by the subsequent anchylosis of bones, which generally remain separate.

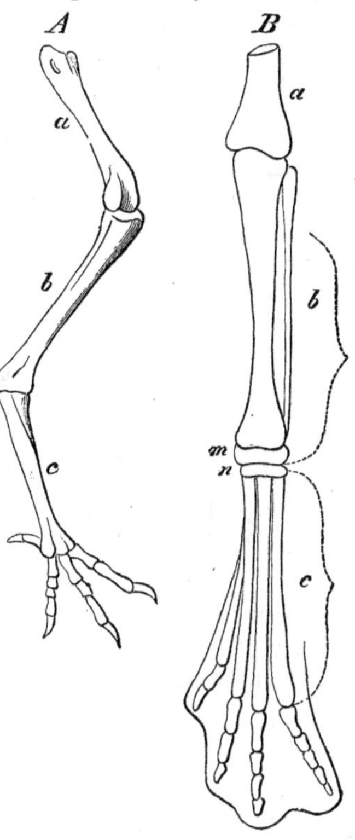

FIG. 1.

The complete leg of the bird (A) shows us at *a*, the femur, or thigh bone, and at *b*, the tibia, or lower leg bone; but instead of the bones of the tarsus and metatarsus, the latter of which affords attachment to the toes, we find only the long bone *c*, and at its lower extremity a small bone supporting the four toes. Earlier writers were content to say that the astragalus (*c*) replaces the tarsus and metatarsus. But this is not the case; for the chick in the egg (B) shows that the bird's leg consists of the thigh, or femur (*a*), and the shank or tibia (*b*), two tarsal (*m n*), and three or four metatarsal bones (*c*), and the toes, or phalanges; that the upper tarsal bone is anchylosed with the tibia, and the lower one with the consolidated metatarsus. Only thus do we obtain a true perception of the fact manifested in A, although the cause of the fact does not as yet appear.

The next example is rather more difficult. Without the history of development, comparative anatomy is incapable of explaining why man possesses three little bones in the auditory apparatus, the bird only one. The history of development shows that out of the material which in man is applied to the formation of the malleus and incus, two other portions of the skull are evolved in the bird, having little or nothing to do with the auditory mechanism. In short, the history of development, which describes the gradual formation of the organism, is at every step a beacon to comparative anatomy. In itself, however, the history of development does not as yet exceed the rank of a merely descriptive branch of erudition.

But if we now perceive how the evolutionary stages of individuals represent series from the lower to the higher, analogous to the various members existing side by side in the same group of animals,—how, for instance, the mammal passes through stages at which the lower vertebrata remain fixed,—a connection, at first sight mysterious, is indicated between the evolution of the individual and the general constitution of the animal world. This connection requires a scientific solution, a reduction to causes, and this all the more urgently because their relations, though as yet hidden, are rendered more probable by a third series of phenomena, the conquest of which is likewise the achievement of natural history. We allude to the record of the primæval world.

Therefore, the knowledge of palæontological facts also forms part of the indispensable basis of our operations. Geology entered the right track forty years ago. We now know that the world was not made backwards, but originated by gradual formations and metamorphoses; we may—nay, we must, infer that, at a definite epoch of refrigeration, life appeared in a natural manner, that is to say, without any incomprehensible act of creation; and during this slow transformation of the earth's crust, we see living beings also gradually increasing, differentiating, and perfecting themselves.

Yet more. As was first convincingly proved in detail by Agassiz, one of the most vehement antagonists of the theory of descent, we behold the palæontological or historical series of organisms in the same sequence as the phases of the development of the individual. There are here vast chasms yet to be filled up by future observation, though in many points we must not altogether despair of success. But that the process of palæontological development is, in general, the one indicated, is disputed only by naturalists, who, like Barrande, years ago anchored themselves to inalterable convictions in science, as in creed, to dogmas.

These groups of facts, thus mutually referring to each other, must be, in some degree, examined by any one desirous of understanding them. In other words, we must first review this vast material, before we turn our attention to the magic spell which sifts and makes it comprehensible. The toil is great, but the reward is glorious! For, as regards the organic world, the craving inherent in the human mind for the knowledge of reasons —the need of causality, is satisfied singly and solely by the doctrine of Descent. As yet we do not regard it as complete; in many special cases it still owes us an answer; but, on the whole, it does as much as any other ingenious theory has done; it interprets by a single principle those great phenomena which without its aid remain a mass of unintelligible miracles. In a word, it raises the knowledge of organic nature to a science. Even now much of mere professional knowledge is wont to style itself science. But as the doctrine of Descent includes all life, it cannot stop on approaching Man. Were we doubtful as to the origin of language, or even forced to admit total ignorance on this point, we could not, from the existence of language, deduce the inapplicability to man of the doctrine of Descent, without, as it seems to us, arbitrarily breaking the chain of ratiocination.

We will now return to the preliminary question already indicated, as to the limits of the investigation of nature. It is the more important, as incompetent judges are wont to assert, that these limits are exceeded. The frivolity of the logic by which such accusations are rendered plausible to the multitude surpasses all licence. We open, for instance, Luthardt's "Apologetic Lectures on the Fundamental Truths of Christianity," ("Apologetische Vorträge über die Grundwahrheiten des Christenthums,") and see how he defends the reality of miracles. "Miracles," he says, "are not even miracles. They do not even repeal the laws of nature; they merely release single occurrences from the dominion of those laws, and place them under the law of a higher will and a higher power. Of this we have many analogies in lower spheres. If my arm hurls a stone into the air, this is contrary to the nature of the stone, and is not an effect of the law of gravitation, but the interposition of a higher power and a higher will, producing effects which are not the effects of the inferior powers. These powers and these laws are not hereby repealed, but still subsist."

Let us pause a moment. To say that it is contrary to the nature of the stone that gravity should be apparently overpowered for a few moments by muscular agency, is physically absurd. The stone remains the same weight, its nature is wholly the same, even while in the motion of projection; and it is utterly unjustifiable and sophistical to prate about muscular force as a higher power opposed to gravity. If the stone weighs two hundred-weight, where is the higher power then?

But when the champion of supernaturalism has misled and prepared his hearers by his worthless analogy, he proceeds: "Thus in the miracle, a higher causality interposes, and evokes an effect which is not the effect of the concatenation of those lower causalities, and yet subsequently submits to these concatenations. But this higher causality ultimately coincides with the highest moral objects of existence. To serve them is nature's highest and most glorious pursuit. Therefore if miracle stands in connection with these objects, if its conditions are moral and not arbitrary, it is not contrary to nature and its purpose, but in the highest sense conformable to it."

Thus as soon as belief in miracle comes into conflict with the investigation of nature, it says: "You overstep your limits, and must here suspend your judgment. It is a question of a higher moral object; the domain of ethics is higher than that of physics, and therefore a higher causality, which physicists have no right to criticise, has suspended the chain of cause and effect with which you naturalists are familiar." This passage[1], in which one of the most learned and honoured champions of the belief in miracle lays down, like a sophist, the limits of the investigation of nature, is, however, among the most moderate of its kind. But our point of view and our logic differ radically from that of antagonists of this description, in one particular, namely, that to us the opposite to knowledge is ignorance, whereas they supplement knowledge by a so-called higher knowledge, and by faith.

While holding by the maxim of Pico della Mirandola, "Philosophy seeks, Theology finds, Religion possesses the Truth,"[2] it is forgotten that there are truths and truths. The subjective visions and sensations of sound by which the mentally diseased are excited and alarmed, are to them a reality, yet a reality quite different to that of the sights and sounds received through the healthy organs of the senses. Philosophy and science seek that truth which is deduced from the palpable connection of things. But the other truths, so often negatived by the former, are generally impalpable, and are incommensurable with scientific truths. We will therefore abide by the words of Goethe:

Whoso has art and science found,
Religion, too, has he;
Who has nor art nor science found,
His should religion be.*

And now, having provisionally averted uncalled-for objections and conflicts with ambiguous ideas, we may quietly consider the limits of natural science. Let us first pause at the address delivered with general approval by the physiologist Dubois-Reymond, at the fiftieth assembly of German Naturalists and Physicians. He made reference to a passage in the classical works of Laplace, in the Introduction to the Theory of Science, which we cannot refrain from quoting in full. The author of the "Mechanism of the Heavens," says: "Present events are connected with the events of the past by a link resting on the obvious principle that a thing cannot begin to exist without a cause which produces it. This maxim, known by the name of the Principle of Sufficient Cause, extends likewise to events with which it is not supposed to come in contact. Even the freest will cannot evoke them without a determining impulse." "We must, therefore, regard the present condition of the universe as the consequence of its former, and the cause of its future, condition. A mind, for a given moment acquainted with all the forces which animate Nature, and the reciprocal relations of the entities of which it is

* Wer Wissenschafft und Kunst besitzt,
Hat auch Religion;
Wer jene beiden nicht besitzt,
Der habe Religion.

composed—possessed, moreover, of powers of comprehension sufficient to submit all these facts to analysis, would be able to reduce to a single formula the motions of the largest heavenly body and of the lightest atom. To such a mind nothing would be uncertain, and the future, like the past, would lie open before it. The human mind in all the perfection which it has been able to give to astronomy, offers but a faint image of such a mind as this." "All efforts of the human intellect in the search for truth tend to approach the mind above portrayed, but will always remain infinitely removed from it."

The Prussian physiologist then quotes the "Thou art like the Spirit whom thou comprehendest" of Faust;* and is of opinion that, in the abstract, the formula of the universe is therefore not impenetrable to the human intellect. But we own we are cordially indifferent to an abstract perfection which never comes to light, and regard the unattainableness of this vague formula of the universe as a very endurable limit to human inquiry. But independently of the dubious consolation of the formula of the universe, we must agree with Dubois-Reymond, when he considers that the limits, before which the highest conceivable intelligence must pause, are also insurmountable to man.

In accordance with the views now prevailing among physicists and biologists, Dubois-Reymond has thus specified the only limit given to the investigation of nature[3]: "The knowledge of natural science, more closely defined above, is no real knowledge. In the attempt to comprehend the constant, to which the mutations in

* Du gleichst dem Geist, den du begreifst.

the material world may be traced back, we stumble on insoluble contradictions. An atom contemplated as a minute, indivisible, inert mass, from which forces emanate, is a chimera. In the impossibility of comprehending the nature of matter and force lies the only limit to the knowledge of natural science."

These propositions require some elucidation. Beyond the subdivision mechanically possible, we must think of substance or matter as consisting of particles ultimately indivisible. Of these atoms, according to the present standpoint of science, we are obliged to admit as many different species as are not chemically reducible to more simple elements. Now there is no doubt that these atoms are, in the actual sense of the word, imaginary, hypothetical quantities; and theory seems to indicate that all matter, in the most different phenomena in the material world, is based on a single species of atom.

Every manual of physics or physiology will show that, in order to understand and calculate the properties of these atoms and their combinations into the ingredients of compound bodies, susceptible of chemical analysis, they are ideally represented under various material forms, spherical, cubical, &c.; furthermore, that in their combinations and co-operations as bodies, they must be contemplated as surrounded by a rarefied atmosphere of an universally diffused ether. But the atom itself, and therefore the nature of matter, is something incomprehensible, unattainable. In these atoms, forces are inherent, which display themselves in attractions and repulsions, and in motion in general. But the final cause of these motions, and how far these motions are, as it were, identical with the existence of the atoms, is likewise included in the incomprehensibility of matter.

"If we pass over this," says Dubois-Reymond again, "the universe is approximately comprehensible. Even the appearance on the earth of life in the abstract does not render it incomprehensible. For life in the abstract, contemplated from the standpoint of the theoretical investigation of nature, is merely the arrangement of molecules in a state of more or less stable equilibrium, and the introduction of an exchange of material, partly by their own elastic force, partly by motion transferred from without. It is a misapprehension to see anything supernatural in this."

This is the point which is usually contested with the greatest vehemence. If all the motions and states of quiescence of the inanimate world can be thoroughly explained, the inexplicable must commence with the basis of life. The imputation cast upon the reasoning powers by this assumption may be formularized as follows, in the question put by another sound and thoughtful physiologist, A. Fick:[4] "Are the characteristics of such a particle, as already explained, applicable and effective during the period of its sojourn in an organism? Thus, for instance, will the motions of a particle of oxygen be affected and altered by a neighbouring particle of hydrogen, in accordance with the same laws, when one or both form part of an organism, as when they are out of it?"

To reply in the negative is to avow the vitalistic conception of life, that is, to take refuge in unknown forces quite extraneous to matter, and to admit that the self-same particle can vary its nature, according to whether it be internal or external to an organism, is, in other words, to affirm a miracle. If this is weighed against the physical view, "which in its perfection reduces every organic process to a problem of pure mechanics," it may be done in the certainly impartial words of the naturalist just quoted: "I am of opinion that the mechanical view of organic life is demonstrated only when all the motions in an organism are shown to be the effects of forces, which at other times also are inherent in the atoms. But similarly I should regard the vitalistic view as proved, if in any case a particular motion actually observed to take place in an organism were shown to be mechanically impossible. At present, neither is to be thought of. Nevertheless, if a decision must be made without full proof, I provisionally profess myself unequivocally in favour of the mechanical view. Not only does it recommend itself *à priori* by its superior probability and simplicity, but the progress of scientific development raises it almost to a certainty. When it is seen how certain phenomena—such as the evolution of animal heat, which it was formerly believed could be explained only by vital force—are now ascribed, even by those who in general assume the existence of a special vital force, to the universally active forces of the material particles, we find ourselves almost forced to the conviction that by degrees all the phenomena of life will become susceptible of mechanical explanation."

For the elucidation of the example just given of animal heat, let us observe that modern physics have learnt to know heat as a peculiar mode of motion. The motion of the hammer as it falls upon the anvil is not lost, but is transformed into the atomic motion of the places struck, a motion, invisible, it is true, but sensible as heat. But likewise the combination of the particle of oxygen introduced into the animal body by the respiration, with the un-oxygenated constituents of the blood, is a motion subject to computation, and manifesting itself as oxydation, combustion, or the evolution of animal heat. This chemical act of combustion keeps the animal steam-engine in motion.

In this way, by the application of mechanical principles, modern physiology has traced to their causes a great number of organic processes, and the phantom of vital force, which formerly reigned paramount over the whole intestinal canal, incited the glandular cells and the muscular fibres to their offices, and glided along the nerves, now scarcely knows where to breed disturbance.

Thus the investigation of nature does not shrink from enrolling life and the processes of life in the world of the comprehensible. We are foiled only at the conception of matter and force. But we are much further advanced than Schopenhauer and his adherents, who for the idea of Force substitute that of Will; for we have analyzed into their several self-conditioned momenta a multitude of processes, which the word "Will," incomprehensible in itself, is supposed to explain in their totality; and much further also than the fashionable philosopher of the day, von Hartman, who regales us with the agency of the "unknown" in the domain of the organic world.

"And yet," Dubois-Reymond thus formulates another limit, "a new incomprehensible appears in the shape of consciousness even in its lowest form, the sensation of desire and aversion. It is, once for all, incomprehensible how, to a mass of molecules of nitrogen, oxygen, hydrogen, carbon, phosphorus, and so on, it can be otherwise than indifferent how they lie or move; here, therefore, is the other limit to the knowledge of natural science. Even the mind imagined by Laplace cannot go beyond this, to say nothing of our own. Whether the two limits to natural science are not, perchance, identical, it is, moreover, impossible to determine."

In these last words the possibility is indicated that consciousness may be an attribute of matter, or may appertain to the nature of the atoms. And we may add, that the attempt has of late been repeatedly made to generalize the sensory process, and to demonstrate it to be the universal characteristic of matter, as by von Zöllner, in his work on the Nature of Comets, which has created such a justifiable sensation. He holds that, if by means of delicately-formed organs of sensation it were possible to observe the molecular motions in a crystal mechanically injured in any part, it could not be unconditionally denied that the motions, hereby excited, take place absolutely without any simultaneous excitement of sensation. We must either renounce the possibility of comprehending the phenomenon of sensation in the organism, or "hypothetically add to the universal attributes of nature, one which would cause the simplest and most elementary operations of nature to be combined, in the same ratio, with a process of sensation."

It might be imagined that reflections of this kind would lead to the delusive abysses of speculation; but if, still speaking only of organisms, we descend from the manifestations elicited by sensations of desire and aversion in the higher consciousness of man and of the superior animals, till we see all reaction to external excitation dwindle into the scarce perceptible motions of the simplest protoplasmic animalculæ, it is evident that there can be no question here of either consciousness or will. We cannot then separate the idea of those sensations of desire and aversion, by which motions are excited, from the elementary attributes of matter, as we are wont to do with regard to the higher animals.[5]

In precisely the same sense, it was said some years ago by one of the most talented investigators of language—Lazarus Geiger, now unfortunately deceased:[6] "But how is it, if further down, below the world of nerves, a sensation should exist which we are not capable of understanding? And it probably must be so. For as a body that we feel could not exist unless it consisted of atoms that we do not feel, and as we could not see a motion were it not accompanied by waves of light which we do not see, neither could a complex living being experience a sensation strong enough for us to feel it also, in consequence of the motion by which it is manifested, if something similar, though far weaker and imperceptible to us, did not occur in the elements, that is to say, in the atoms. If we only consider that we are as little capable of knowing that the falling stone feels nothing, as that it does feel; it is fully open to us to decide, in accordance with the greatest probability, that the world is susceptible of explanation."

We have examined the limits which the investigation of nature has prescribed for itself. The organic world, far from rearing itself before us as an incomprehensible entity, invites us to fathom its nature, and promises to reflect fresh light upon the inanimate world.

We must now pass in review a great portion of animate nature, and shall then arrive at the same conclusion as the linguistic inquirer, to whom—we again quote his words—"it became, on historic grounds, incontrovertibly certain that man has risen from a lower, an animal grade."

II.

The Animal World in its Present State.

IN order to approach the doctrine of Descent, and to prepare for its necessity, we purpose next to pass in review a main part of its object,—the present condition of the animal world in its general outlines. Organisms, as every one may see, are distinguished from animate bodies by a certain mutability of existence; a sequence and alternation of phenomena, combined with constant absorption and expulsion of matter. These changes, which are ultimately molecular motions, and are therefore calculable, definable, and susceptible of investigation, take place in particles in a state of saturation—that is to say, soaked in water and aqueous fluids; and this peculiar, yet purely mechanical condition, suffices for the explanation and comprehension of many of the necessary phenomena of life. Experience shows that this capacity for saturation, and this mobility, essentially characterize the combinations of carbon; and the sum of these motions and displacements, of which a great part has already been susceptible of mathematically certain investigation, is termed Life.

Now it is impossible to resist the impression that there are simple and composite, lower and higher, living beings; and we likewise feel, more strongly than words will express, a certain antithesis between the plant and the animal. Poetically regarded, the plant is the passive organism as described by Rückert:

> "I am the garden flower
> And meekly bide the hour,
> The guise, with which you come
> Within my narrow room." *

The antithesis of the passive, quiescent plant and the pugnacious active animal diminishes, however, as we descend in the scale of both kingdoms. The more highly developed animal evinces its animal nature by the vivacity with which it reacts to external influences and excitations. In the lower animals the phenomena of life assume a more vegetal character, and in many groups of lower beings, which Haeckel has recently comprised under the name Protista, we see the processes of metamorphosis of tissue, nutrition, and reproduction taking place, indeed, but in a manner so simple and undifferentiated, that we too must attribute to these beings a neutral position betwixt plants and animals. We gain the conviction that the roots of the vegetal and animal kingdoms are not completely sundered, but, to continue the simile, merge imperceptibly into each other by means of a connective tissue. In this intermediate kingdom the much derided "primordial slime" (Urschleim) of the natural philosophers has regained its honourable position. Many thousand cubic miles of the sea-bottom consist of a slime or mud composed in part of manifestly earthy inorganic portions, in part of

* "Ich bin die Blum' im Garten
Und muss in Demuth warten,
Wann und auf welche Weise
Du trittst in meine Kreise."

peculiarly formed chalk corpuscles, still perhaps ambiguous in their nature (the Coccoliths and Rhabdoliths), and finally, which is the main point, of an albuminous substance which is alive.

This living slime, the so-called Bathybius, does not even exhibit individuality, or the definiteness of a separate existence; it resembles the shapeless mineral substances, each particle of which bears the characteristics of the whole.

The conception of an organism as a being composed of various parts, with various offices or functions, and appearing under a definite form gradually developed, is in our day so inherent and intuitive, that it is only with great exertion that we are able to accommodate ourselves to the idea of a living mass either absolutely formless and undefined, or defined arbitrarily and accidentally. Let any one, who either cannot or will not do this, pause for a moment to contemplate another simple being—for instance, Haeckel's "Protamoeba." A small albuminous mass increases by the absorption of nutriment, and by the appropriation of matter, until it reaches a certain circumference, and then propagates itself by spontaneous fission into two equal parts. To our means of observation, these and similar beings are the simplest organisms devoid of organs. While accentuating the limits of research as restricted by inadequate means of observation, we maintain the validity of Rollet's retort,[7] that our reason cannot properly admit such homogeneous organisms, performing all the functions of life solely by means of their atomic constitution; that we are dealing with the still utterly unknown structure of the molecules formed by the aggregation of atoms; and that if Brücke says, "Apart from the molecular structure, we must also ascribe to living cells another structure of a different order of complexity, and this is what we denote by organization," we must likewise ascribe this yet unknown combination to the Monera of Haeckel.

But independently of this complexity of the molecular structure, it is of extreme importance to the investigation of animate nature to have become acquainted with bodies which present the simplest structure to the assisted eye, and to anatomical research. The substance which characterizes them is found again in plants as well as in animals; and plants and animals must now be regarded as two classes of organisms, in which the processes of self-preservation and reproduction have, in different ways, assumed the character of a higher complexity and development, by the differentiation of the originally homogeneous substance into various morphological structures and organs.

As we shall have another opportunity of expressing an opinion in regard to the beginnings of animal life, and its points of contact with *protista* and plants, we shall transfer ourselves from the dubious boundary line into the midst of the animal kingdom, in order to master our subject by sifting and arranging it.

The first impression of infinite variety is succeeded by another, that there are lower and higher animals. On this point complete harmony prevails. For if, from teleological considerations, invalid in our eyes, the nature of every creature were said to be perfect, that is, in correspondence with its purpose or idea, every one takes it for granted and self-evident that a standard of excellence exists, without taking account of the scale by which it rises or sinks. This standard will, however, soon be made manifest by the comparison of a lower with a higher animal. Let us select the fresh-water polype and the bee.

The little animal, several lines in length, which in our waters usually lives adhering to a plant, is a hollow cylinder, of which the body-wall is formed of two layers of cells, a layer of muscles, and a supporting membrane, which gives consistency to the whole, and may be compared to a skeleton. The mouth is surrounded by arms of similar construction, and varying in number from four to six. The surface of the body is studded with numerous little stinging vesicles, which by their contact stun any smaller animalculæ straying within the reach of the polype, and render them an easy prey. This is, in a few words, the construction of the animal. It possesses no arterial system, no special respiratory apparatus; the functions of the nerves and the sensory organs are performed by the individual parts of the surface. Reproduction is usually effected by the budding of gemmules, which fall off at maturity, but occasionally also by the produce of very simple sexual organs.

On the other hand, hours do not suffice to describe the structure of a bee. Even externally, its body, which possesses so highly complicated a structure, promises a rich development of the interior. The manducatory apparatus can be rendered comprehensible only by comparison with the oral organs of the whole insect world. The various divisions of the alimentary canal are each provided with special glands. The rich psychical life, all the actions which imply intelligence, calculation, and perception of external situation, are rendered possible by a highly developed nervous system, and the marvellously complex sensory organs combined with it, of which the eyes are especially remarkable. Independently of the generative organs, consisting of manifold parts of greater or less importance, the history of the multiplication and development of the bee demands a study of itself.

The function, and therewith the rank and value, of the bee's body seem to us higher than that of the polype in proportion as it is more complex. The superior complexity and variety of the parts is anatomically evident, and similarly the higher phase of the life. The superior energy of the existence, the functional capacity and perfection of the bee as contrasted with the feebleness of the polype, is obviously a result, or more correctly an expression, of the greater mechanical and physiological division of labour. In one animal, as in the other, life is spent in the function of self-preservation and the maintenance of the species, or reproduction ; in both, the cycle of phenomena is limited, unbroken ; but the means of execution are very different, and therefore the general effect is different. In the variety and correlation of the organs destined for the different manifestations of life, we have a standard for the rank of the animals. This rank has a twofold character, general and special. In other words, the position of an animal in the system is defined, first, by the general attributes, which it has in common with the forms harmonizing with it in the main characters of their organization ; and, secondly, by the more special characteristics, which place the animal in its own rank and station among its own immediate kindred.

Some insight into this classification of the animal kingdom is naturally indispensable to any one, who wishes to test and understand its reasons, and to render an account of it is an essential part of our task.

Since Cuvier's reconstruction of Zoology in the early part of this century, our science has been familiarized with the expression "type," or "fundamental form," introduced, long before, by Buffon. Cuvier, by extensive dissections and comparisons, first proved that animals were not, as people were formerly inclined to suppose, made on a last or shaped upon a block; but that they fall into several great divisions, in each of which expression is given to a peculiar constitution, arrangement, and distribution of the organs; in short, to a peculiar style. The sum of these characteristic peculiarities, as well as the whole of the species united in it, was termed a "type." Various views, it is true, even now prevail as to the extent of several of these types or families, as we will already term them; but if we disregard the dubious, and in many ways suspicious, existences, generally comprised under the name of primordial animals, there is a general agreement as to the following number, but less as to the sequence of the animal types, than as to those groups, each of which has its peculiar physiognomy and special characteristic structure.

The class Cœlenterata includes the Polypes and Medusæ, and in the closest connection with it stands the interesting class of the Spongiadæ, especially instructive as affording direct evidence of the doctrine of Descent. The organs of these animals are nearly always arranged radially round an axis, passing through the dorsal and ventral pole. The cavity, which in most other animals—for instance, in man—is termed the abdominal cavity, the space between the intestinal wall and the abdominal parietes, is deficient in them ; but, on the other hand, from the stomach proceed in general various kinds of tubes and branchia, which to a certain extent replace the abdominal cavity. Fig. 2 represents a

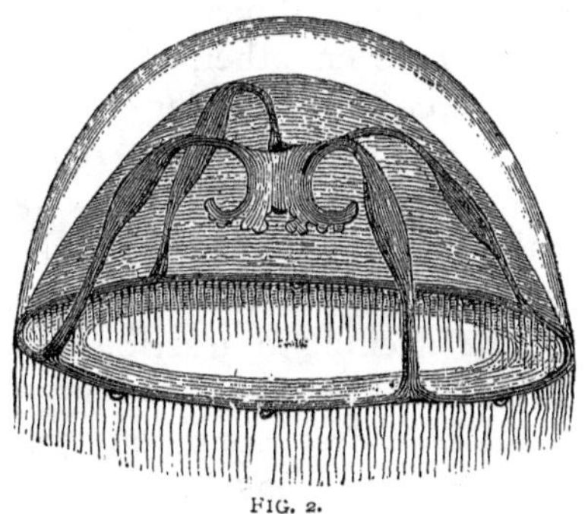

FIG. 2.

Medusa, Tiaropsis Diadema, after Agassiz. The darkly-shaded organs form the so-called cœlenteric apparatus.

Of the Echinoderms, the reader is probably acquainted, at least with the star-fish (Asterias) and the sea-urchin (Echinus), of which the general form is likewise usually radiate. Besides a peculiar chalky deposit, or greater or less calcification of the skin covering, a system of water-canals forms a characteristic of this family. With these are connected the rows of suckers, which, by protrusion and retraction, serve as organs of locomotion. On account of the radiate structure prevailing among the Echinoderms, Medusæ, and Polypes,

Cuvier believed them to be more nearly related, and introduced them altogether, under the name of Radiata. This similarity, however, is only superficial, for whilst, on the one hand, anatomy discloses the great difference of the Cœlenterata and Echinodermata, the history of evolution still more decidedly banishes the Echinoderm from this position, and connects them more closely with the next division.

In this, that of the Vermes, the systematizer of the old school finds his real difficulty; in so many ways do they deviate from each other, so great is the distance between the lower and the higher forms; and after deducting the distinctive marks of orders, so little remains as a common character, so variegated is the host of smaller scattered groups, and even of single species, which demand admittance to the system of the Vermes. If we attempt to describe their typical nature in a few words, it must be something like this: The Vermes are more or less elongated, symmetric animals, which possess no actual legs, but effect their locomotion by means of a muscular system, closely combined with the integuments, which frequently become an actual muscular cylinder. To this we will add, that the perplexities and difficulties in reference to points of classification are transformed into sources of knowledge for the adherent of the doctrine of Descent.

The relations of the previous family with the type of the Articulata is so conspicuous, that the "kinship" of the two was never questioned, even by the older zoologists. The very name of one, the highest division of the Vermes, that is, of the Annelids, or segmented worms, indicate this connection. This distinctive mark of the Crustacea, Arachnida, Myriopoda, and Insecta, is that their bodies are constructed of sharply-defined rings or segments, the legs, antennæ and mandibles likewise sharing in this segmented character. A faithful expression of this segmentation is afforded by the nervous system, which lies, ladder-like on the ventral side, that is, beneath the intestinal canal, nearly encircling the gullet with its anterior loop. The display of segmentation is favoured by a deposit of horny substance, which gives a skeleton-like stiffness to the integuments.

The direct reverse is shown in the integuments of the Mollusca, our mussels, snails, and cuttle-fish. For although so many are supplied with protecting scales and shells, these are mere excretions from the actual skin, which remains soft, and characteristically moist and slimy, owing to the secretions of numerous glands contained in it, and has an inclination to lay itself in folds, and form a mantle-like investment to the body. The body therefore remains more or less clumsy; it possesses none of the grace of the Articulata, and especially of the insect; it is destitute of segmentation, and this deficiency is likewise evinced in the nervous system. This consists only of a ring, encircling the œsophagus, and a few smaller ganglia.

We shall most readily come to an understanding as to the Vertebrata, the family with which man is inseparably united. The essential part is the vertebral column, that portion of the internal and persistently bony or cartilaginous skeleton, in which the main portion of the nervous system is contained.

It is thus established that the systematic classification of the animal kingdom is based on certain prominent characteristics of form and internal structure; and it is very easy to select from every type forms in which the distinctive marks, comprised in the systematic diagnosis, may be displayed in full perfection. But this is immediately succeeded by a further observation, that of gradations within the type. When we previously compared the polype and the bee, and were obliged to assign to each a very different rank, a portion of this difference of grade is certainly due to the difference of the family; but the forms united by family characteristics likewise diverge widely from each other, and the systematist speaks of lower and higher classes within every type, of lower and higher orders within every class.

Reason is compelled to this by the same considerations which forced themselves upon us in the comparison of the polype and the bee. Why does the mussel stand lower than the snail? Because it does not possess a head, because its nervous system is not so concentrated and so voluminous, because its sensory organs are more defective. In one, as in the other, the structural material is present in quantities sufficient for the completion of the type; but in the snail it is more developed, and the single circumstance of the integration of various parts to form the head confers a higher dignity upon the snail. It is needless to illustrate this gradation within the families by further examples; the most superficial comparison of a fish with a bird or a mammal, of one of the parasitic crustacea with a crayfish or an insect, shows, as the older zoology represented it, that in the actual forms the ground plan, or "ideal types," find very diversified expression.

A further result of this descriptive inquiry is the

tree-like grouping of the members of the same family. The reciprocal relations of the various families cannot be represented in a simple line; though in former days more importance was attributed to the general indications of the relative value of the types. On the other hand, descriptive zoology had long been compelled to devise tables of affinity for the systematic subdivisions, descending even to species according to the criterion of anatomical perfection; and these found expression only in diagrams of highly ramified trees. Branches appeared which terminated after a brief extension; others are greatly elongated with numerous side branches; in every branch characteristic phenomena and series are made manifest.

Let us attempt it with the Vertebrata, for example. Even with the fishes we fall into great perplexity; which to place at the end as being the highest. But take which we will, the sharks or our teleostei, the amphibians cannot be annexed in a direct line, nor does the elongated branch line of the latter merge, as might be imagined, into the reptiles. The birds, on their side, offer a sharp contrast to the mammals, and this separation and divergence extend to all the subdivisions. We must figuratively represent family branches, clusters of genera, and tufts of species, which latter ramify into sub-species and varieties. With this representation of the tree-like distribution of the system, we shall gladly revert to the comparison of the members of different types, with reference to their functional value. The bee in itself is manifestly a far more complex organism than the lowest fish-like animal, the lancelet; and in these two we compare a low form of a high type, and a high form of a low type. By varying and combining comparisons of this sort, and taking account of the points of connection between the various types, to which we shall immediately refer, the figure of the systematic trees completes itself into one vast tree, of which the main branches are represented by the types.

Had the systematizers of the old school been familiar with the construction of plants and animals, they would have first established the diagnoses and distinctive characters, and then called to life the types and their species; for their chief torment has been, that the diagnoses are liable to so many exceptions, and that the characters of the fundamental forms are without any absolute value. Roughly and generally speaking, polypes are radiate in form, but not a few are bilateral, or symmetric on two sides. Most snails possess well-marked mantle-folds, but we can scarcely speak of the testa of many thoroughly worm-like slugs.

Head and skull seem an inalienable mark of the vertebrata, yet the lancelet has no such head, but merely an anterior end. Nevertheless, it may be objected, it has a vertebral column ; yet this, the special badge of nobility of the vertebrate animals, like the auditory apparatus, and the notochord, is, even if only transiently, a possession of the Ascidians, a class of animals which in their mature condition do not bear the remotest resemblance to the Vertebrata. When we become aware of these deviations from so-called laws of form and structure, seemingly well established, we are prepared for a manifest failure of the system, in regard to connecting forms, and forms of uncertain position in the system.

If the result of the systematic sifting and arrangement within the individual types can be comprised in diagrams of trees, forms intermediate to the members of the types, classes, orders, &c., follow as a matter of course. For if the figure be correct, every ramification of the branches must include species diverging very slightly from the species standing in the lowest portions of the bough from which it branches off. And thus all systematizing, in fact, amounted to the insertion of the right intermediate forms between each two forms deviating from each other in a higher degree; nay, in some cases, intermediate forms were sought where none exist. The older zoology always regarded the duck-mole (Ornithorhynchus) as the mammal most nearly allied to the birds, though the cause of the bird-like appearance of the lowest mammal known, is by no means to be sought in a direct relationship, but in a remote cousinhood.

But we must draw attention, not to these connecting forms, which natural history assumes as perfectly self-evident, but to those which are, as it were, inconvenient to systematic description, and threaten to render illusory the groundwork so laboriously gained. There are some fish-like animals, the Dipnoi, (Lepidosirens and their congeners) with the characters of Amphibians. The Infusoria possess many characteristics of the so-called primordial animals, but in other ways they differ from them, and point to the lowest Turbellaria. A minute animal inhabiting our seas in countless multitudes, *i.e.* the Sagitta, is neither a true annelid nor a legitimate mollusc. The class of the Radiata fits neither into the system of the actual Annulosa, nor into that of the true Articulata, yet provision must be made for it in the system; and any one who clings to the types as ideal and inalterable fundamental forms, falls into sad perplexity how to dispose of his Radiata.

Example after example might be thus accumulated to show that the rigid partitions of the system are scarcely raised before they are again broken down in every direction; and this in direct ratio with the increase of special science. As before said, descriptive natural history necessarily gained this experience. It then spoke of exceptions and deviations, without being able to adduce any reason why the classes and types should be able to break through their limits, and indeed most frequently without feeling any need of accounting for the failure of the rigid system.

III.

The Phenomena of Reproduction in the Animal World.

THE faculty of giving existence to new life is part of the evidence of life. A crystal does not reproduce itself, it can only be resolved into its elementary constituents; and in the natural course of things, or in an artificial manner, these may be induced to form another crystalline combination. But this is not that continuity of reproduction which links individual to individual, is not procreation wrapped in a cloud of mystery. Herein, it seems, consists a stubborn opposition. Yet, if the distinction between animate and inanimate nature has been recognized as one not entirely absolute; especially if the possibility, nay even the necessity, has been perceived of the primordial generation or parentless origin of the lowest organic beings from inorganic matter (of which more hereafter), and if the nature of nutrition and growth is understood to be entirely dependent on the power of obtaining material,—the mystery of reproduction henceforth disappears. Generation is no longer a mystical event; and the origin of an organism in or from an organism, the emission or development of innumerable germs, may, like the origin of a new crystal, be analyzed into the motions of elements, as yet accessible only to the eye of imagination. By this we mean to say that in the province of reproduction the limits of inquiry are neither narrow nor peculiar. We will therefore now proceed to describe the process of reproduction and development in the animal kingdom.

If, as must be generally admitted, the most essential characteristics are common to the highest and the lowest life,—and it is only the complexity of the vital processes, together with the variety of the parts by which they are performed, that give rise to graduated diversities,—it will, of course, be in the simplest organisms that we shall most readily recognize the nature of these vital processes.

The simplest beings, discovered by Haeckel, such as the Protamœba, those minute albuminous masses of sarcode, increase to a certain extent. Why these dimensions should vary only within definite narrow limits, and why, on attaining a certain extent, the molecules should gravitate into two halves, we do not know; at any rate it is an affair of relations of cohesion, theoretically susceptible of computation. It is enough that at a certain size the coherence of the parts is loosened in a central zone, the individual becomes faithless to its name, and divides into two halves, of which each from the moment of separation begins an individual life, while from the commencement of the fission preparations were being made for their self-dependence. This is the simplest case of reproduction, a multiplication by division. Frequently, however, it does not stop at bisection; the motion of the minute constituents, which causes the fission, proceeds in such a manner that the halves are again divided, and the quarters yet again, the whole being thus divided into a greater number of portions, and the parent-creature is resolved into a swarm of off-shoots.

This multiplication by mere division of the mass presupposes that the organism thus reproducing itself possesses no high complexity. The bisection of a beetle or a bird is inconceivable as a means of propagation. Yet Stein's valuable observations on the reproductive process of the Infusoria, make us acquainted with organisms standing far above these simple so-called Monera, of which the subdivisions undergo a series of profound metamorphoses, before separating as self-dependent individuals. This transformation, combined with fission, leads to reproduction by gemmation.

As the fission of these low organisms depends on the attainment of a certain limit of growth conditional on adequate nourishment, the case now more frequently occurs that the individual discharges the superfluity of material obtained at a definite part of the body, and forms a bud or gemmule. We are already acquainted with reproduction by gemmation in the simplest organism, the cell; for all healing and cicatrization in higher beings, even to the re-integration of the mutilated limbs of amphibians, is effected only by the reproduction by fission and gemmation of the elementary morphological constituents. But it lies in the nature of the process of gemmation, that it should extend far higher than fission in the scale of organisms; it is the origination of a new being from one already existing, the latter, meanwhile, preserving its individuality wholly or for the greater part, and yet being able to transfer to the progeny its own characteristics in their full integrity.

The simplest case of gemmation is where the parent

animal produces one or more gemmules similar to itself, capable in their turn of producing similar gemmules. Of this, every collection of corals gives numerous examples, and shows how the diversified appearance of the several genera of coral depends merely on minor modifications of this mode of reproduction. Yet single corals exist in which, on careful comparison, not only may accidental deviations be already discerned, but regularly recurring variations between parent and progeny, as Semper has recently shown in Madrepores and Fungiform corals. This brings us to the highly-important phenomenon of Alternate Generation, which we must elucidate by a few examples before entering upon the nature of sexual reproduction.

Figure 3 shows in A a polype-shaped being with cruciform tentacles, on which its discoverer, Dujardin, bestowed the generic name of Cross-polype, or Stauridium. This animal, growing like a polype upon a stalk, forms above its lower cross, gemmules which make their appearance as spherical balls, gradually assume a bell-like shape, and detach themselves on attaining the structure and form of a Medusa or sea-nettle. The Medusa (termed Cladonema Radiatum, Fig. 3 B) is thus the offspring of its utterly dissimilar parent, the Stauridium ; it reproduces itself in the sexual method, and from its eggs proceed Stauridia. The two generations thus alternate; the cross-polype is an intermediate generation in the development of the Medusa, so that the sexual generation never originates directly from its egg.

In the tape-worm, we have an illustration of the same process, only in a somewhat more complicated form. It is known that from the intestinal canal of individuals afflicted with tape-worm, issue so-called somites or segments of the tape-worm. These somites are usually filled with such an extraordinary number of ova that they seem like mere packets of eggs. It appears, how-

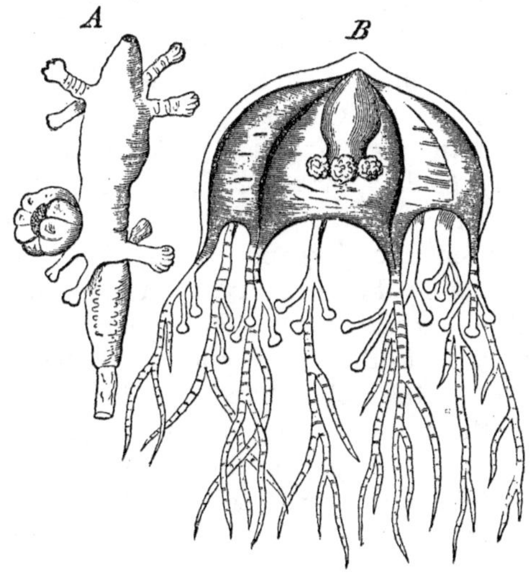

FIG. 3.

ever, from the evolutionary history of the tape-worm, and its relations with other annulosa, namely with leeches and Turbellaria, that notwithstanding their incompleteness and deficiency of organs, these somites are equivalent to sexually mature individuals; or, according to Haeckel's definition, are endowed with personality. If the tape-worm now comported itself like most other animals, somites would be directly developed from its eggs. But to this there is a very circuitous proceeding.

If the egg of a tape-worm, by chance and good luck, strays into a congenial stomach,—for example, the egg of the human tape-worm, Tœnia solium, into the stomach of a pig, the embryo wanders out of the stomach in which it quitted the egg, and makes its way into the muscles, where it swells out into a sort of cyst. This cyst is the first intermediate generation. It produces a peg-shaped gemmule, which, however, fails of its object as long as the "bladder worm," or "Gargol," remains in the flesh of the pig. It is only when this comes, raw or imperfectly cooked, into the human stomach, that the time has arrived for the release of the pupa. It emerges from its parent the cyst, and the pupa, in which we now recognize the head and thorax of the tape-worm imago, represents a second intermediate generation. Its productiveness is forthwith displayed; it becomes elongated, and as its ribbon-like form increases, shooting out from the posterior portion of the cervix, the more distinctly marked become the transverse stripes and "somites;" in other words, the individuals of the third or sexual generation.

In the evolutionary cycles just discussed, there is an alternation of asexual and sexual reproduction; and before examining some other cases of asexual multiplication, we must make ourselves acquainted with the facts of sexual reproduction.

The characteristic of this is, that it requires for the generation of the new individual the union of two different products or morphological elements, the ovum and the sperm. The ovum is always, in the first instance, a simple cell, of which the nucleus is termed the germinal vesicle, and the nucleole the germinal spot.

In many animals it is provided with a sheath or membrane of its own ; in others it remains naked, and in that case frequently displays the remarkable movements of protoplasm. The germ-cells of different classes of animals vary considerably in their microscopic dimensions ; nevertheless, in the whole animal kingdom, from the sponges and polypes up to the mammals inclusive of man, they are essentially similar. Nor do non-essential differences appear until the primitive germ-cell is more abundantly provided with yelk and albumen, and has surrounded itself with a specially thick and perforated shell, as in insects and fishes, or with a peculiarly formed sheath, in the shape of a double concave lens, as, for instance, in some Turbellaria. As a rule, the ova are formed in special organs, the ovaries. The other sexual element, the sperm, contains, as its peculiar active constituents, the spermatozoa (fig. 4 *s*), which consist of a pointed, elliptic, or occasionally of a hook-shaped, head, and a thread-like body. As long as the sperm is capable of fecundation, the filamentous appendage performs serpentine movements, and the development of the spermatozoa from cells, as well as the comparison of their movements with the vibrating movements of ciliated and flagellate cells, enable us to recognize them also as modified cell structures.

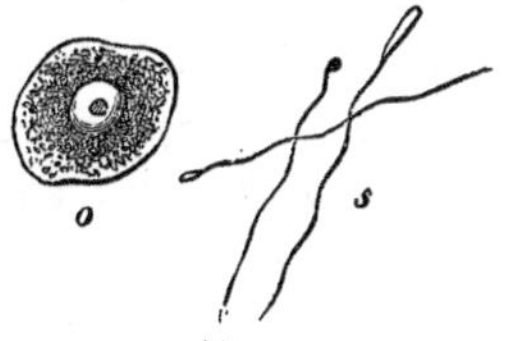

FIG. 4.

The vehement dispute of last century between Evolutionists and Epigenists has now a merely historical interest. The former maintained that either in the ovum or in the sperm-corpuscle the whole future organism was prefigured in all its parts, and that it hence required only the development of the infinitely minute organs already existing. The others, who carried off the victory, saw in the ovum the yet undifferentiated material which subsequent to fecundation had still to be transformed into the various morphological elements and organs. But it is scarcely twenty years ago since the process of fecundation was discovered, and since it was proved that at least one sperm corpuscle, and, as a rule, several or many, must penetrate into the interior of the ovum and unite materially with its substance in order to produce an effectual fecundation.

The course of our demonstration obliges us to place sexual in sharp contrast with asexual genesis. But here, again, recent times have produced a series of equalizing and conciliatory observations which must not be neglected by us, bent as we are on tracing the antecedents of the doctrine of evolution, and demonstrating the transition taking place throughout organic Nature. In the cases of alternate generation selected above, the generations which do not produce ova and spermatozoa, reproduce themselves by external gemmation. Now, there is manifestly no great physiological difference if the deposition of the material from which the progeny is formed takes place, not externally, but in and by special internal organs. One of the most familiar examples occurs in the evolutionary cycle or alternate generation of the genus Distoma of the Entozoa. In the ventral cavity of one larval generation arise cell-spheres, or germs, which develope into the second generation—the Cercaria.

Great excitement was likewise aroused by the discovery of the germ-formation of the larvæ of a dipterous insect (Cecidomyia, Miastor). In the ventral cavity of the maggots of these flies arises a second generation of maggots, of which the origin was primarily attributed to a simple germ-formation, until it was shown that these germs proceed from the situation of the sexual glands (which in many insects are developed at a very early stage), and must therefore be regarded as unfertilized ova. The second generation of maggots lives at the expense of its parent, consumes its fatty substance, and afterwards destroys the other organs; while of the pelican-like parent nothing finally remains but the skin, as a protecting cover to the offspring, which very soon emerges.

Without mentioning other cases in which it may be questionable whether germs or unfertilized ova attain development, we will point out a few of those in which development, without fecundation, is established with complete certainty. The queen bee, partly from the natural course of its life, partly from various accidents in which fecundation could not take place, lays regularly a number of unfertilized eggs, from which issue drones, or male individuals; or if exceptionally eggs are laid by workers, which are imperfectly developed female bees not susceptible of fecundation, these eggs likewise produce drones only. Von Siebold's highly interesting experiments on the reproduction of a wasp (Polistes Gallica), have shown that the hybernating fertilized females, who found a new colony in the spring, deposit eggs whence issue female individuals, and occasionally males. This virgin generation then produces eggs from which males are developed. With various butterflies, on the contrary, the unfertilized eggs produce females only; and it is the same with several of the lower crustaceans.

We will now revert to the consideration of the evolutionary processes displayed in sexual reproduction after fecundation has taken place. Development invariably commences with a process of cell-formation, the bifurcation or formation of the germinal membrane, after the completion of which, instead of the one primitive cell, a large number of cells are usually in existence, as the material for the distribution and construction of the embryo. Ova developing parthenogenetically, without fecundation, likewise commence their development by this multiplication of cells; and even the ova of animals, in which development never takes place without previous fecundation, exhibit an incomplete bifurcation, if not fertilized at a certain stage of maturity. This process, it is true, has been as yet demonstrated only in the ova of the frog and the domestic fowl; but these cases are sufficient to divest the bifurcation of the character of an independent phenomenon, exclusively restricted to sexual reproduction.

Even before the appearance of C. E. von Baer's really classical and fundamental work on the "Evolutionary History of Animals" (Entwickelungsgeschichte der Thiere),[8] the view, founded on incomplete observations, had become established, that in the various stages of their development the higher animals passed through the forms of the lower ones. In this, natural philosophy did not confine itself to the limits of the types; and hence did not pause at the hypothesis that the mammalian embryo was successively a fish, an amphibian, and in a certain sense, by a particular gradual evolution of the organs, a bird also, but made the embryo likewise repeat and surpass the lower types. To this false tendency, acting on vague analogies, a stop was put by the great naturalist just named. He showed that a number of coincidences might, indeed, be demonstrated between the embryo of the higher and the permanent form of the lower animals, but that this resemblance rested essentially on the fact that in the embryo of the higher animal the differentiation of the general fundamental mass had not yet set in, and that in the progress of development it passes through stages which are permanent in the series of inferior animals.

On the other hand, he positively repudiated the assertion that the embryos of the higher types actually pass through forms permanent in the lower ones. He says that the type of each animal seems from the first to fix itself in the embryo, and to regulate its whole development. As regards the vertebrate animals in particular, the further we go back in the history of their development, the more do we find the embryos alike, both on the whole and in the individual parts. "Only gradually do the characters appear which mark the greater, and later those which mark the smaller divisions of the Vertebrata. Thus from the general type the special one is evolved."

Von Baer thus held that the analogy consisted only in the embryonic states of the various animal forms; but he was obliged to go beyond the circle of the types, and he thought it probable that among all embryos of vertebrate, as well as invertebrate animals, developed from a true ovum, there is a conformity in the condition of the germ at a period when the type has not yet manifested itself. This led him to the question, "Whether, at the beginning of development, all animals are not essentially alike, and whether a common primordial form does not exist for all?" "It might," he finally thinks, "be maintained, not without reason, that the simple cyst-like form is the common fundamental form from which all animals are developed, not merely in idea, but historically."

When the barrier which it was formerly thought necessary to erect between asexual multiplication and multiplication caused by fecundation had been recognized as non-existent, and it was perceived that all development amounts to the multiplication and metamorphosis of the primitive germ or egg-cell, the cell was necessarily regarded, in the acceptation of the older investigators, as the common fundamental form. But although the descriptive history of evolution does not go back to this elementary organism, and considers even the bifurcation as merely a preparation for actual development, at any rate the earliest rudimentary larval conditions of different types may be compared with each other.

The discoveries of the last ten years with reference to this subject are so numerous, and such striking analogies have been advanced, that we must needs go much further than, at that time, was possible for Von Baer. It is not merely a question of those general analogies in the segregation of tissues from an indifferent rudimentary mass, but of homologies in the distribution, form, and composition of the embryos and larvæ, of which the after effects are of profound importance to the later and actual typical impress. With this object, let us consider the larva of a calcareous sponge at the stage which Haeckel has designated as the Gastrula phase.

The diagram gives the section of a larva of this description, which at this period is nothing more than a stomach provided with an orifice (fig. 5 *o*); its wall con-

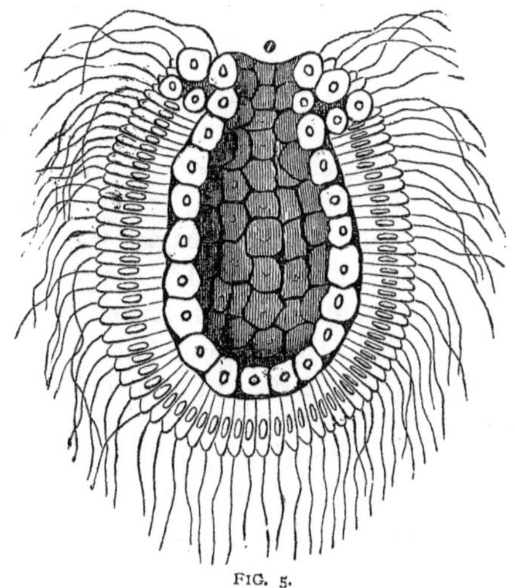

FIG. 5.

sists of two strata, or layers of cells. The cells of the external stratum are distinguished from those of the inner one by their elongated form, and the possession of filaments serving as organs of locomotion. All subsequent development and differentiation, certainly not very important in the sponges, may be traced to modifications of these two membranes; the external membrane (Ectoderm, or Exoderm) and the internal membrane (Entoderm). And this phase of the ciliated larva, with its twofold strata, its primitive ventral cavity and mouth, recurs in the Cœlenterata, with slight variations in the Echinoderms, in some of the Annulosa, in the Sagitta, the Ascidians, and the Lancelet. From the analogy of all these animals, and especially of the last, we shall be able hereafter to derive important inductions.

But if no weight be attached to the presence of these filaments of the external layer, which is, moreover, justified by the relation of the filament to the cell, and if it be acknowledged as the essential significance of the larval arrangement, that from its two laminæ the collective organs derive their origin, then to the animals above enumerated must be added, not only almost the whole of the Articulata, but likewise the remainder of the Vertebrata, as in them, immediately after the appearance of the primitive striæ, follows their separation into two cell-layers, or membranes. Respecting the derivation of the third or middle germinal lamina, and the share of the two primitive laminæ in its formation, observers are not agreed.

Only from this point does the development of the great animal groups take various directions, and it is the immortal merit of Von Baer to have fixed these types of development, independently of the fundamental forms, established by Cuvier on zoological and anatomical considerations, and he thereby laid a far deeper foundation for the existence of these types. We will illustrate our meaning by two examples.

When the ovum of the articulate animal has surrounded itself with a germinal membrane, a portion of it thickens into a long germinal stria, resembling an elongated ellipse. This is the rudiment of the ventral side of the future animal. A groove then divides it into the two germinal laminæ, and transverse striæ next make their appearance, the indications of the so-called primordial segments. The symmetrical disposition of the organs, and the integration of the body out of consecutive segments, is herewith initiated. All further development emanates from these primordial segments, which are the standard of the Annelids or higher Vermes; while in the Articulata, projections and appendages of these segments develop into feelers, manducatory apparatus and legs, and by their heterogeneous integration in the regions of the head, and of the middle and posterior portions of the body, give rise to the vast variety within the type. In each particular case we see what is special emanate from what is more homogeneous and undifferentiated, and this is likewise corroborated by the more advanced phase portrayed in the diagram (fig. 6). It represents the embryo of the great black-beetle (Hydrophilus piceus) on its ventral side. The antennæ (*f*), the three pair of oral appendages (*m*), and the three pair of legs, are as yet little distinguished. In the further course of development, the lateral portions grow towards the back, in the centre of which they finally meet. As compared with the Vertebrata, it may hence be said

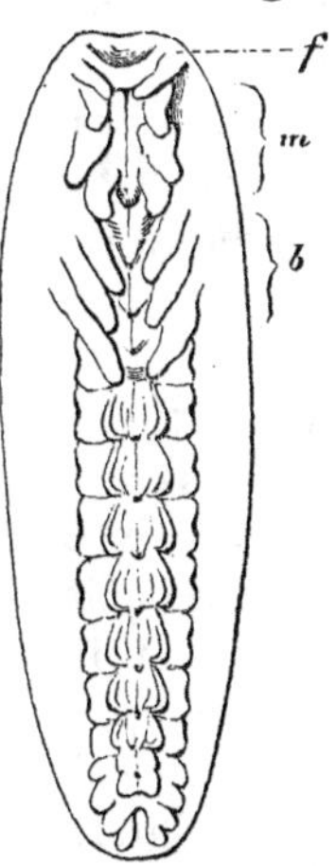

FIG. 6.

that the Articulata have their navel on their backs. Conversely, it is the characteristic of the evolutionary type of the Vertebrata that the position of the germ corresponds with the dorsal side of the animal. The formation of the dorsal groove, which subsequently closes to form the canal of the spinal cord, as it is gradually enveloped in a sheath growing from below, is followed by the formation of transverse plates, the pre-vertebral plates. The side plates lying outside of these grow towards the ventral side, and finally merge in the navel. The position of the actual vertebral column, consisting of separate vertebræ, is always originally occupied by a cartilaginous band, the notochord (chorda dorsalis), and, as from this axis, the germinal matter transforms itself into a tube above as well as below,—into the spinal marrow with its sheath, and the ventral cavity with the intestinal canal,—Von Baer considered this mode of development as bi-symmetrical. The development of the Articulata he regards as simply symmetrical, and the development of the Molluscs he designated as massive. The justification of this is that the elongation produced by segmentation and the repetition of similar parts and sections of the body implicit in segmentation generally,—the metameric formation, as it is termed by Haeckel,—is totally foreign to the Molluscs.

We must now again repeat, that somewhat extensive observations of the evolutionary forms of different animals lead at once to the belief that the embryos and evolutionary phases of higher animals are transiently more closely related to the complete and definitive conditions of the lower animal-forms, at least of the same family; whence arose the fixed idea that the embryo of the higher animals passes through the forms of the lower animals. When natural philosophy, more especially in Germany, had elaborated this doctrine in a rather fantastical manner, and had proclaimed that Man was the sum of all animals, in structure, as well as in development, "the doctrine," says Von Baer, "of the uniformity of individual metamorphosis with the vague metamorphoses of the whole animal kingdom necessarily acquired great weight, when, by Rathke's brilliant discovery, germinal fissures were demonstrated in the embryos of mammals and of birds, and the appropriate vessels were soon afterwards actually revealed."

The exaggerations and false inferences drawn from general analogies, and the vague ideas of types hovering above the whole, and regulating individual development, were wittily chastised by Von Baer.

"To convince ourselves that a doubt as to this doctrine is not utterly groundless, let us imagine that the birds had studied the history of their development, and that it was they who now investigated the structure of the mature mammal and of man. Might not their physiological manuals teach as follows?—'These quadrupeds and bipeds have much embryonic resemblance, for their cranial bones are separate; like ourselves during the first four or five days of hatching, they are without a beak; their extremities are tolerably like each other, as are ours for about the same time; not a single true feather is to be found on their bodies, only thin feather-shafts, so that, even in the nest, we are more advanced than they ever become; their bones are not very hard, and like ours, in our youth, contain no air at all; they are utterly destitute of air-sacs, and their lungs, like ours in early infancy, are not full-grown; a crop is completely wanting; gullet and gizzard are, more or less, merged in a sac, all conditions very transitory in us, and, in most, the nails are awkwardly broad, as with us before breaking the shell; the bats, which appear the most perfect, are alone able to fly; not the others. And these mammals which, so long after birth, are unable to find their own food, and never rise from the ground, fancy themselves more highly organised than we?'"

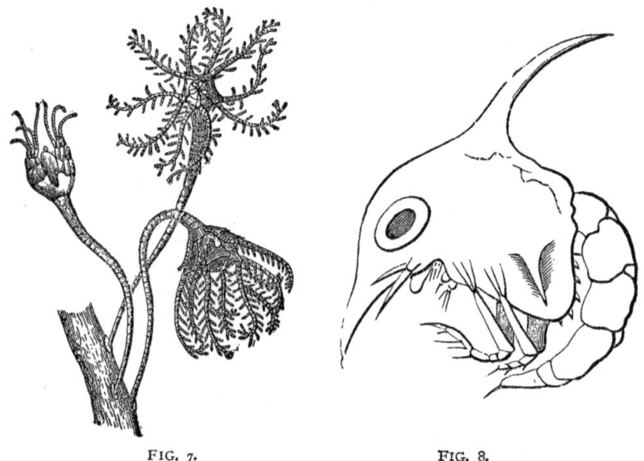

FIG. 7. FIG. 8.

Nevertheless, there remains the fact of the parallelism of individual development with the systematic series to which the individual belongs; and, among thousands of examples, we will select some of the most accessible and convincing. Polypes have always been placed systematically below the Medusæ; in the development of many

Medusæ (comp. Fig. 3; p. 43), a polype-like condition is interposed. The crinoid (Comatula), very common in the Mediterranean, is in its mature condition freely movable. This definitive development is, however, preceded by a sessile stage (Fig. 7), during which the body is attached to a stalk. During the larval period the animal resembles the permanently sessile genera, which, by all systematic rules, and by their geological position, occupy a lower rank in the series of echinoderms. The crabs, or anourous crustacea, are raised by sundry characteristics above their long-tailed congeners, among which is the fresh-water crayfish. In the course of development they pass through the long-tailed stage, as is shown in the larva (Fig. 8). It is by the abortion of the tail, which is employed by the long-tailed species as a natatory organ, that they become more fitted for running, and some of them for terrestrial life, as they are, in a measure, released from a burden.

One of the systematic series included in the Vertebrata, leads through the reptiles to the birds. Now, if, in the physiological reflections which Von Baer put into their beaks, the birds, as will appear later, were mistaken in boasting of their feathery garb in contrast to mammals and to man, they have, nevertheless, carried it a stage further than the reptiles, for the scale is the embryonic rudiment of the feather. Likewise, the tarso-metatarsal joint of the embryonic bird, with which we are already conversant (p. 9), and which is distinguished from the ankle-joint of mammals and of man, by its lying not between the leg and the tarsus, but in the tarsus itself, remains, as a definitive condition in the reptile, in the embryonic condition which in the bird it rapidly passes through. Although mammals are never actual fish, there is much that is fish-like in the embryonic phases of their organs; the embryonic fissures in the thorax correspond with the germinal branchial fissures; the formation of the brain may be traced to the complete brain of the lampreys and the sharks, &c.

In order to refute the doctrine that the embryo passes through the whole animal kingdom, Von Baer was content to prove that it never changes from one type to another. He repudiated the other, and more probable part of this theory, that is, that, at least within the types, the higher groups, in their embryonic phases, repeated the permanent forms of the lower ones, by terming it a question of mere analogies. The embryo, as it is gradually perfected by progressive histological and morphological differentiation, necessarily accords, *in this respect*, with less developed animals in proportion to its youth. "It is, therefore, very natural that the embryo of the mammal should be more like that of the fish, than the embryo of the fish is like the mammal. Now, if the fish be regarded merely as a less perfect mammal (and this is an unfounded hypothesis), the mammal must be considered as a more highly developed fish; and, in that case, it is quite logical to say that the embryo of the vertebrate animal is originally a fish." [10]

We have been somewhat faithless to our intention of confining ourselves in this chapter to facts only. The facts are too apt to provoke reflections, and we have, moreover, repeated these reflections merely as historical facts; we must now inquire whether they are really capable of satisfying us. I think not. It is by no means a merely histological and morphological differentiation which causes the resemblance of the higher incomplete, to the lower complete forms. To limit ourselves to one example: it is quite incomprehensible why the ear-bones of the mammal should be developed, by the circuitous process of the formation of germinal fissures, if it were a mere question of histological and morphological differentiation. This explanation fails also with regard to the whole class of the phenomena of purposeless and abortive organs, and, finally, the "evolutionary type" itself as it rules the groups and regulates individual development, still remains without an explanation.

IV.

The Animal World in its Historical and Palæontological Development.

It is so easy to observe that the earth's crust, from the deepest valleys to the highest mountain top contains innumerable animal remains, that even antiquity could not fail to notice it. But some two thousand years passed by before a correct knowledge was attained of the relations of these remains to the present world. Some thought they were sports of nature, products of creative power leading to no special object, but in a certain measure to be regarded as exercises preliminary to the actual creation of life; others considered the fossils as remains of living creatures, indeed, but of such as still existed, and which had been destroyed by overflows and subsequent withdrawals of the sea. The legend of the universal deluge, especially, derived great support from this second opinion. Only when, at the end of last century, the stratification of the earth's crust was revealed to science, after the outlines of a history of the solar system and of a special history of the earth or geology had been indicated by Kant and Laplace, only then arose the possibility and necessity of a real palæontology, or knowledge of prehistoric life. At the beginning of this century it was discovered that the fossils corresponding with the stratification of the earth's crust follow each other in regular sequence, and that in this sequence they differ from the present creation, as they do from each other.

We must make ourselves acquainted with the order of succession of these strata. They are the shelves in which the vegetable and animal remains lie stored. To arrange them was certainly possible only by taking the organisms which they contained as guides or clues. We, however, shall take this arrangement as our data, and, with the object we have in view, we shall naturally consider only those strata and rocks in which fossils—using this word in its widest interpretation—are or might be contained, those, namely, which are proved to be sedimentary, *i.e.* aqueous deposits. Our information is limited to a great part of Europe, numerous districts of America, and scattered points of the rest of the world.

The following table gives the the arrangement of the sedimentary strata from above downwards :—

1. Alluvium.
2. Diluvium.
3. Tertiary formation.
   - Pliocene.
   - Miocene.
   - Eocene.
4. Cretaceous formation.

Sinon.	=	White Chalk and Chalk Marl	of English Series. See p. 307 of Page's Advanced Text Book.
Turon.	=	Part of the Chalk Marl	
Kinoman.	=	Upper Greensand	

   - Gault.
   - Neocoman (Wealden).
5. Jurassic formation or Oolite.
   - Upper White Jura (Malm).
   - Middle Brown Jura (Dogger).
   - Lower Black Jura (Lias).

6. Triassic formation or New Red Sandstone.
   Keuper or Variegated Marls.
   Muschelkalk.
   Variegated Sandstone.
7. Permian formation or Dyas.
   Zechstein (Magnesian Limestone or Dolomitic Conglomerate).
   Rothliegendes or Red Conglomerate.
8. Carboniferous formation.
   Coal Measures.
   Millstone Grit.
   Mountain Limestone.
9. Devonian formation.
10. Silurian formation.
11. Cambrian formation.
12. Laurentian formation.

Although we are not writing on geology, a short explanation of these strata will be requisite, as their mutual relations also throw light on the nature and distribution of the contemporaneous organisms. All displacements of earth which we now see occurring by means of rain, rivers, sea and other natural forces which have taken place in historic times, in short, in the so-called Present, such as the great delta deposits, and the moraine formations of our glaciers, are ascribed to the *Alluvium.*

It was formerly supposed that its limits might be distinguished from the *Diluvium* by the appearance of man, but as it is now, and always has been impossible to affirm anything positive respecting that epoch, and as, although a portion of the organisms of which the remains occur in the Diluvial strata is extinct, much more still exists, these two formations are inseparably intermingled.

To the Diluvium belong the vast mud deposits of the great rivers, alternating with sand banks, the clay and loess formations caused by the removal of the soil by the drainage of the glaciers and the floods of running water, which at one time increased periodically to a degree truly colossal. The diluvial period, as it seems, includes, both in Europe and America, a repeated glacification of countries and vast portions of the world, of which the present state of Greenland may now give some idea.

The period of the series of strata, comprised under the name of the tertiary formation, may be regarded as that during which, at least, the skeleton of the present continents finally attained its integral configuration. Within its limits fall the erection and upheaval of the great mountain chains, the Cordilleras, Alps, Himalayas, and others ; the outlines of the continents were, meanwhile, in constant movement. This phenomenon, however, persists throughout all formations, and, as the geological characteristic of the tertiary formation, more stress should be laid on the separation of the earth's surface into climatic zones, approximating to the zones of the present age. The names of the subdivisions are intended to indicate the relation of the animals then living to those of our world, as it was supposed that in the eocene the first animals identical with present species were to be found, more in the miocene, and, yet more, in the pliocene.

To the chalk formation belong rocks of very various kinds, which can be reduced to one great geological period by means of their contents. If the quartzose sandstone of Saxon Switzerland represents this formation in the centre of Germany, it is from the white chalk of England and Northern France that it took its name. In America, the sandstone has been in a great measure ground down into sand, and in other places the strata are purely chalky or marly. But the vagueness of the limitations of strata in situation, and still more in time, may be estimated by the fact that we are fully justified in speaking of the chalk formation now going on, as is shown by the investigations of Carpenter and W. Thompson on the constitution of the deep sea-bottom of the Atlantic. To the early chalk period belongs a great fresh-water deposit, and likewise the Wealden, a formation of peat and bog occasioned by upheavals, which contains a number of remains of fresh-water and terrestrial animals, besides a peculiar sort of coal.

The oolitic strata appear more definite, mostly lying regularly over each other in distinct deposits, more rarely, as in the Alps, raised up by later dislocations. The rocks themselves, betray that the depositions took place in wide seas, for the most part calm or deep, and this is rendered a certainty by the scanty vegetal remains and the far more abundant animal remains which they contain. In the apparently very sharp limitation of the oolitic formation, both above and below, the older geology found a main prop for the assertion, that comparatively quiet periods of long duration alternated with catastrophes destroying and re-creating everything. To avoid any misapprehension we must, however, add that the oolitic period already possessed vast and highly integrated continents, as it will likewise be seen that during this era the higher terrestrial animals made their appearance.

The characters shown by the three great divisions of the triassic formation are very various, especially as they are developed in Germany. The German portion, judging by its influxes, must be regarded as a formation of strands and bays; its more highly integrated equivalent in the Alps as a huge oceanic deposit. The Muschelkalk (which is missing in England), with its layers of rock salt and rich remains of oceanic organisms, is likewise a marine formation. Of the origin of the stratified variegated sandstone, so-called from its varied colouring, with its clays, marls, and frequent vast enclosures of gypsum, we obtain some idea from our present formations of sandy shores and dunes. Like these, the deposition of the variegated sandstone afforded but scanty opportunities of enclosing animal and vegetal remains, but very notable footprints have been preserved, such as might now be formed and preserved, if the marks imprinted on the damp sand were filled up with fine clayey particles torn by a storm from some adjacent shore, and subdivided in the sea.

As the diversified appearance of the superimposed planes of antediluvian plants and animals of course depends essentially on the nature of their former abodes, and as the nature of the individual districts of each plane must then, as now, have influenced the character of the organisms by which it was inhabited, we will indicate the causes which thus affect life in its form and manifold variety. In order to complete our view of the origin of the Earth's crust, and the dependence of the organic on the configuration of the inorganic world, we will leave a geologist, Credner, to describe the relations of the dyassic and carboniferous formations: "In regions where the carboniferous (coal) formation is typically developed, it consists of a series of stratifications, the lower one chalky (mountain limestone), the middle one conglomerated or arenaceous (millstone grit), and the upper one carboniferous (coal measures); hence a marine, a littoral and a marsh or fresh-water formation. It is easy to imagine the cause of this phenomenon; it depends on the secular elevation of the primæval sea bottom, on which was deposited first the marine mountain limestone; secondly, as it rose to the surface, the shingle and coarse sand of the shore; and finally, on persistent elevation, the products of marshes, lagunes, and estuaries. If it now happened that some portions of the infant continent covered with the latter (that is to say, with the productive carboniferous strata), were seized with an opposite movement, and therefore sank, there would be deposited on the surface now again gradually becoming the bed of the sea, precisely similar forms, only in inverse order to that which occurred during the period of elevation.

And, in fact, this phenomenon is exhibited by those portions of the earth's surface which shortly after the formation of the coal measures again sank below the sea. In Germany and England the productive coal measures are followed by a sandstone and conglomerate, therefore a littoral formation, exactly like the quartzose sandstone and millstone grit which underlies them; and above this a limestone, dolomite and gypsum formation, corresponding to the mountain limestone, the lowest member of the carboniferous system. On account of the division which is displayed in profound palæontological and petrographical diversities, the formation thus developed and composed is designated as the Dyas. The separate phases of this cycle of occurrences, by which the carboniferous and Dyassic formations were evolved, are accordingly (reading from above downwards):

5. Deep Sea.	Marine forms.	Mountain Limestone.	Marine animals.	Magnesian Limestone.	
4. Subsidence beneath the Sea.	Littoral forms.	Conglomerate and Sandstone.		Red Sandstone.	Dyas.
3. Quiescence.	Freshwater and Marsh forms.	Coal measures.	Terrestrial plants.	Coal measures, Red Sandstone, Coal fields.	
2. Upheaval above the Sea.	Littoral forms.	Conglomerate and Sandstone.		Millstone grit.	Carboniferous formation.
1. Deep Sea.	Marine forms.	Mountain Limestone.	Marine animals.	Carboniferous Limestone.	

From this account it is also manifest that in cases of incomplete elevation, such as took place in North America, the formation of the middle period is either disturbed or totally omitted, and that it may depend on local causes and the duration of the oscillations if, as in the Russian Permian formations, corresponding to the German Dyas, the boundaries of the subdivisions are more or less obliterated.

The two series of strata beneath the mountain limestone, and reaching the depth of more than 3000 and 6000 metres, the Devonian and Silurian formations, are the lowest, and therefore the first which clearly bear the mark of their origin as marine deposits. Both groups were formerly comprised under the name of Transition rocks, or Graywacke formation. In them also sandy, clayey, and chalky rocks alternate with one another, already exhibiting modifications of a local nature, from which, towards the carboniferous period, issued the first beginnings of continental upheaval.

The granite, gneiss and slate, which as primary rocks, or primitive formations, originated before the Silurian rocks, are for the most part sediments of hot or very warm primæval seas, which have undergone manifold internal changes from pressure and heat. Till recently, they were likewise termed the Azoic group, as containing no vestiges of life, when the discovery of the Eozoon and its unlimited occurrence in the Laurentian strata of Canada, proved that the required conclusion to the series had actually taken place.

With this Eozoon we begin the enumeration of the antediluvian animals from below upwards. The remains of this creature consist of a more or less irregular system of chambers with cretaceous walls, of which the interior is filled with serpentine or pyroxene. It was attempted to deny the organic origin of this cretaceous testa, which may best be compared to the shells of the Foraminifera. But renewed researches have substantiated that although in the great mass of the Eozoon rocks occurring in vast strata, metamorphosis has rendered it nearly, if not quite, impossible to recognize the true nature of the body, pieces here and there occur with the chambering so distinctly marked, and a tubular structure peculiar to the Foraminifera, which exclude any other interpretation than that of a living being resembling the low Foraminifera. This is of great significance, as the profusion of life met with in the Silurian and Devonian strata presupposes an immeasurably long antecedent period during which life had already existed and gradually increased to the multitudes of the Silurian era. We discover in it but scanty remains of marine plants, and only marine animals; but these are so heterogeneous and varied in form, that they alone would oblige us to infer the existence of coasts, shallow or deep oceanic regions, and a number of geographical conditions on which we see the variety and extent of animal life to be dependent. Besides numerous forms of corals more nearly allied to still existing families, we find the quite peculiar group of Graptolites (fig. 9), which, although not actual polypes, might be ranged next to the so-called Medusa-polypes, and thus justify the inference that preparation was being made for the appearance of the higher forms of the Cœlenterata, the Medusæ.

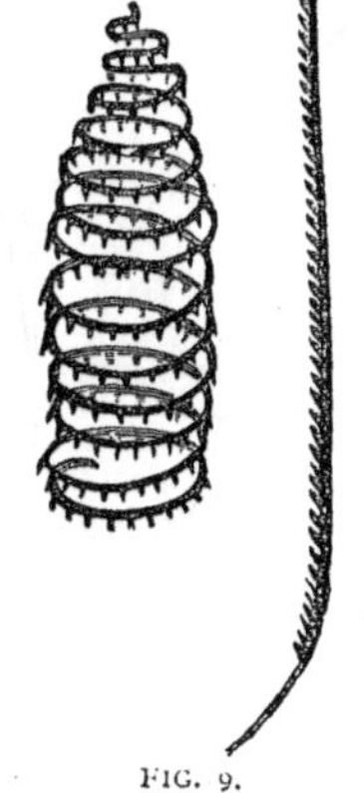

FIG. 9.

The Articulata are represented by the Trilobites (fig. 10, Trilobites remipes), a crab-like form which recalls the present group of the Lamellibranchiata, but has not hitherto admitted of any closer definition, as in none of the many thousand specimens examined, of the forms (about 2000) known in the Silurian and Devonian strata, have the legs been preserved. In these three-lobed crabs, the head, trunk, and tail distinctly appear, as well as the threefold transverse division. The two composite eyes already indicate a high grade of organization. The power of rolling themselves up, which they have in common with several of the crabs now inhabiting shallow waters and coasts, and likewise their general habit, allow us to infer that they also were denizens of coasts.

The Molluscs were mainly represented by Brachiopoda and Cephalopoda. However, as Bivalves and Gaster-

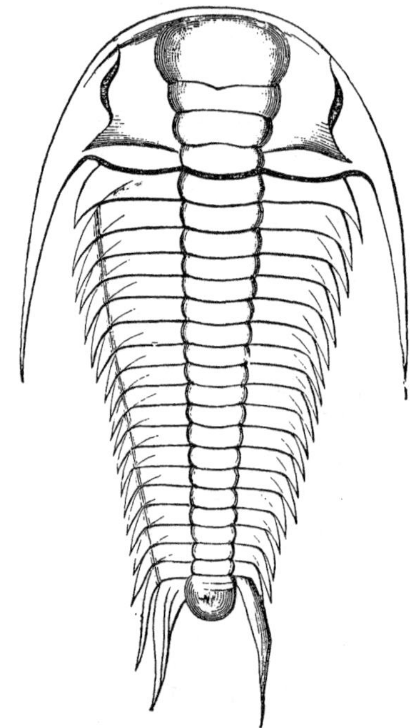

FIG. 10.

opoda were also in existence, the appearance of this, the most ancient molluscous fauna known, differs from the present one only in its numerical proportions, and in the circumstance, certainly very important, that of the Cephalopoda the Nautilus alone is found. The Brachiopoda soon attain to their highest development, and have lingered on till now in a greatly reduced state. Among the Conchifera, the Dimyariæ take the lead in the course of the later period; and with regard to the Gasteropods, we will merely observe that they constantly increase in internal complexity and variety as they approach more recent periods, and that the terrestrial and fresh-water species are occasionally found in the carboniferous formation, though in number and variety they belong primarily to the Tertiary era. To the Cephalopoda we must return again. Of the Vertebrata in the Silurian strata we know only the remains of peculiar Fishes whose kindred must be sought among the sharks and rays.

In the period of the Devonian or upper Transition rocks, the surface of the earth had assumed, at least in places, a more smiling appearance. Here begins the first record of terrestrial plants. As to the character of the fauna, the rapid decrease of the Trilobites is worthy of notice, and the appearance of the important genus of the Cephalopoda, Clymenia, subsequently replaced by the Ammonites. Above all, we must note the increased abundance of fish which still form the sole representatives of the Vertebrata, and held undisputed sway in the seas of that period. Besides the sharks, there are the mailed Ganoids. It is true, the fish, the hinder part of which is here portrayed (Fig. 11, Palæoniscus), belongs only to the upper Coal and Zechstein formation; but it is necessary even now to point out the characteristics of the true Ganoids which floundered about the Silurian seas in somewhat extraordinary forms. Agassiz terms them Placoids, from the rhombic scales, provided with a layer of enamel highly favourable to preservation, and covering the whole surface in oblique rows.

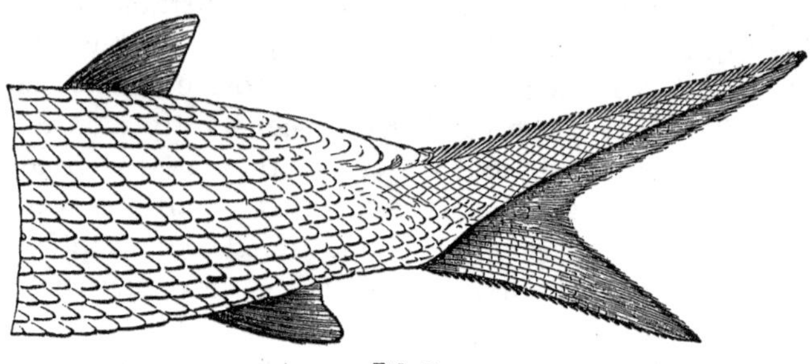

FIG. 11.

The vertebral column, as in the sharks, enters the upper flap of the tail and renders it strikingly unsymmetrical. The Ganoids are, as comparative anatomy has proved with certainty, a development of the shark-like fishes, if not decidedly of a higher grade. The Ganoids, therefore, presuppose the shark.

The carboniferous period owes its name to the enormous accumulation occurring in its midst, of the remains of terrestrial plants, fern-like Calamites, and more especially of Sigillaria and Lepidodendra, standing between vascular Cryptogams and Conifers. They formed tropical bog-forests, such as Franz Unger some years ago attempted to restore in an ingenious composition. In these steaming primæval forests, differing from the early beginnings of antecedent periods by their extent and luxuriance, new phases of animal life become manifest—scorpions, myriapods, and insects—in other words, air-breathing Articulata, and likewise the first air-breathing Vertebrata. The latter, the Cheirotheria, or Labyrinthodonta (colossal Batrachians) possess pre-eminently amphibian characters, and exhibit, for example, several important characteristics of the Batrachian skull, whereas their skin-covering recalls the scale-armour of the Saurians. Thus we find characters combined which are subsequently divided among different groups. There are also traces of huge sea-lizards. But here, and likewise in the magnesian limestone formation, these amphibian-like animals still keep in the background amid the profusion of Ganoids, which especially characterizes some of the strata of the magnesian limestone formation, the Kupferschiefer, or cupriferous marl formation. For the sake of classification, the Zechstein is not unfitly supposed to conclude a great period of organic development: the series of formations from the Silurian to the end of the Zechstein is termed palæazoic; and those which follow, the Trias, Oolite, and Cretaceous formations, are summed up as mesozoic.

The Trilobites, the mailed Ganoids, and others have now disappeared, and the enormous development of reptile life stamps this middle period. The Trias as yet possesses no true Teleostei. The Labyrinthodonta still predominate; while the Archæosauros and the Proterosaurus, which had already appeared in the Dyas, are replaced by more numerous forms approximating to the true reptiles. One single discovery in the upper member of the Trias—the teeth of a predatory marsupial—has supplied us with the most ancient traces of a mammal. It might be inferred, even from the petrographic character of the oolitic strata, that this era must have been, on the whole, far more favourable to the development of animal life than the more perturbed Triassic period, or that at least a more copious preservation of organic remains might be expected, for the oolitic strata are mostly depositions which have taken place without disturbance.

And so it proves. The Placoids and Ganoids hitherto predominating in the ocean almost without a foe, now found overwhelming enemies in the true sea-lizards, or Enaliosaurians, especially the Ichthyosaura and Plesiosaura. The head is like a lizard or a crocodile, the vertebral column fish-like, and, as Gegenbauer has shown, the extremities also recall the simpler classes of sharks. Their coprolites likewise allow us to infer with full certainty, a very peculiar construction of the middle portion of the intestinal canal. They possessed a spiral intestine like that of the sharks and their congeners. These animals are therefore noteworthy, not only on account of their striking external appearance and the part they play in nature's household, but, like the Labyrinthodonton, as mongrel and connecting forms of reptiles and of fish.

In addition to these animals, we must distinguish among the marine fauna the Ammonites, which now appear in vast masses, and the Nautili, the second chief form of the ancient Cephalopods, the study of which has recently promised to contribute essentially to the decision of the most important points in our science. In combination with them, the Belemnites abound with their multitudinous species, originating in the Trias. They are proved to be the predecessors of dibranchiate Cephalopoda, which now predominate. On the chalk plains of Eichstadt and Solnhofen, belonging to the White Jura, are also preserved impressions, resembling drawings, of Medusæ, which show that even at that time this class had reached the state in which it still exists.

The terrestrial fauna of the Jurassic period is likewise enriched by new forms and groups. We find the first true crocodiles, tortoises, and the most remarkable variation of the Sauroid type, the winged lizard or Pterodactyl. It is evident from their well-preserved skeletons that the wing membrane was stretched, as in the bat, between the posterior and anterior extremities. Behind, it extended to the foot, while in front, it obtained a corresponding addition by the elongation of the little finger. A first and only bird has likewise been found in the well-known resting-places of the Pterodactyls, in the lithographic slates of Solnhofen in Bavaria (Archæopterix lithographica). The most remarkable peculiarity of this bird, recognizable by the most minute impression of its feathers, is the long tail, bordered by two rows of rigid feathers. The head is unfortunately crushed beyond recognition. The inferior order of Mammals already mentioned, the Marsupials, were also present, as is shown by the enclosures of the middle Oolite of England and the upper Oolite of the Purbeck strata.

The ornithic animals of the chalk, are more remarkable intermediate forms than the Archæopteryx, and these by their hour-glass-shaped vertebrate bodies are directly connected with the sea-lizards of the Jura, and also possess teeth; this may, however, be the case with the Archæopteryx also. We shall return later to these creatures, which fill up a void hitherto painfully sensible. During this new period the Ammonites were most abundant, and then became extinct, after going through a stage of degenerate forms which may be observed in the

Turrilites, Scaphites, Baculites, and others are considered. The prime of the great sea-lizards is also past, but the marshes of the Wealden period harboured new forms of colossal land-lizards. The long-tailed cray-fishes are joined by the true crabs, the most highly developed forms of the class. In the Oolite and Chalk also occur the chief of the sea-urchin-like Echinoderms. As yet we have not mentioned the class of Echinodermata, in order that we might here point out in conjunction several of the more important phases of their geological occurrence. Desor,* a distinguished judge of this class, has lately examined how in this large group of Echinæ the progress of organization is gradually manifested, on which occasion he was induced to make some general reflections on the principle of progression, as applied to the Echinoderms, probably known to all our readers in their representatives the star-fish and sea-urchins. If articulate, as well as vertebrate, animals attain a higher grade of development by the differentiation of the consecutive segments of the body, the superior unity, and therewith higher perfection, of the Echinoderm's body is evinced when the spines, or so-called antimera, give way to the unity of the whole.

The more distinct these elements are, that is to say, the more independent they remain, the lower is, not only the articulate animal, but also the Echinoderm. Accordingly, the star-fish, and to some extent the feather-stars, stone-lilies, or crinoids, occupy the lowest rank. But here, unluckily, palæontological tradition likewise abandons us. Only so much is certain, that in the older fossiliferous strata both divisions are abundantly repre-

* Bulletin de la Société des Sciences Naturelles de Neufchâtel, IX. 2.

sented. A highly remarkable and important intermediate form is also known, found in the upper Silurian strata of Dudley (Eucladia Johnsoni), the more important as but few transitional forms between one order and another have been hitherto discovered. The relation of the star-fish to the sea-urchins is still indistinct. On the other hand, the bridge from the stone-lilies to the sea-urchins is tolerably apparent. The true Crinoids are sessile, and with them are connected, in the carboniferous formation, the no longer sessile Cystoids and Blastoids, with which are associated the Tessellæ, more resembling the sea-urchins. Now the Dyas and Trias are still poor in true Echinæ ; the Jura, on the contrary, very rich ; and in this great period the extraordinarily heterogeneous transformations of the Echinæ are slowly accomplished, and may be traced, step by step, from the Lias, the earliest oolitic formation, to the coral limestone. At first the Cidaridæ predominate ; they are joined in the Oolite by the Echinoconidæ and Cassidulidæ. In the upper layers of the Jura, the sharper separation of the species becomes characteristic.

Desor shows how this development, accompanied by temporary quiescence, is connected with the nature of the sea-bottom at the time. "The law of progress," he says, "is displayed in the circumstance that it is the lowest of the Echinæ, the Regularæ and Endocyclicæ, which primarily appear, first in the form of the Tessellæ, then as Cidaridæ ; while the most perfect Spatangæ, with the most distinctly marked bilateral form, make their appearance last of all. Between these extremes we find a host of genera and species distinguished from one another by mere shades, so that of two allied genera it is often difficult, nay, impossible, to state which is the more perfect. Progression is only to be shown collectively; in the concrete case it can rarely be demonstrated."

The Échinæ still predominate in the chalk. Recent discoveries of analogous animals, with soft and flexible persistent integuments, confirm what was theoretically extremely probable, that from them proceeded the highest existing order of the Holothuria or Sea-cucumbers; and thus the division of Echinoderms conforms to the universal experience of the ascent from the lower and undifferentiated to the higher forms.

With the Tertiary period dawns the state of things now existing. Palms and arboraceous plants characterize the vegetation. The animal world has likewise remained essentially the same from the earliest sections of the Tertiary period until now, as we shall more elaborately set forth in the chapter on Geographical Distribution. In the most ancient formations the Fishes, in the middle the Reptiles, were conspicuous in the world of life as the representatives of the highest development; now when the continents, not indeed without sundry local oscillations, are approximating to their present configuration, the impress of the Mammalia becomes predominant. Under the influence of elevations and depressions, of several glacial periods, and the more sharply defined limits of the climatic zones, frequent displacements occurred in the vegetal and animal world, accompanied by differentiation and further development. As we have already mentioned, the course of our inquiries will bring us back to this subject.

At the time when geologists believed in the rigid partition of the earth's periods of development and the sharply separated succession of the evidence in its favour, that is to say, of the systems of stratification, the fixed conception of a fossil was, that whatever had lived before the appearance of man on the threshold of the Alluvial period was fossil. It has been proved that the existence of man is far more ancient; that species and races which surrounded the cradle of mankind have become extinct; hence that they, like the Mammoth, for example, are fossil to us only, and not to our diluvial forefathers; while many other animal forms which existed before man have been preserved till now. On the whole, from the Tertiary period forwards, the herbivorous Mammals precede the Carnivora. The monkeys appear only shortly before man.

Notwithstanding many gaps in the palæontological record, the progress of development is manifest in the organic world, including the vegetal kingdom. No fossil animal controverts the system. On the contrary, the most varied adjustments and accommodations are afforded by the antediluvian animals. If, for instance, the present Pachyderms are sharply distinguished from the Ruminants, an unbroken bridge between them is established by the extinct forms. If the present time shows us only single scattered genera of the Edentata, the Diluvial period exhibits a considerable number under far more heterogeneous forms. Thus in the types as in the divisions of the classes, the system advances from the older to the more recent periods; while the more ancient groups gradually increase and then diminish, as newer, more perfectly or specifically integrated forms, are interposed. The former either vanish entirely or outlast the more recent periods, and continue in scanty remnants down to the present day. The formations mostly have their characteristic organisms, but almost everywhere the connecting links have been exhibited. Everything conduces to show that it is a question of evolution, not revolution. Wherever there seems to be a sudden break, the case is the same as in the revolutions of human history, in which likewise only reforms long-prepared, and practically necessary, come to a rapid issue.

If we sum up the result of the comparison of fossil with living animal life, we are first of all struck by the accordance between the grades succeeding one another in the order of time, and the members now ranged side by side in the system. Secondly, when this is confirmed, the parallelism between the geological succession of animals and the grades of the individual development of present animals follows as a matter of course. Agassiz, in his great work on fossil fishes, pointed out this fact with irresistible force, and confirmed it in his later writings by renewed, valuable, and convincing observations on the investigations of the development and growth of corals. The same examples which served in the preceding chapter to illustrate the parallelism of individual development with the systematic stages, may be repeated here; though many newer and very striking instances have been brought to light by the special researches of the last ten years. To express this relation, Agassiz introduced the term "embryonic types," or "embryonic representatives." Thus the stalked stone-lilies are the embryonic types of the present genus Comatula; the most ancient Echinæ are the embryonic representatives of the higher families of the Clypeastræ and Spatangæ; the Mastodon, on account of its persistent molar teeth, is the embryonic type of the elephant, which only transitorily possesses such teeth. If the term implies nothing further than the vague assertion of "the working of the same creative Mind through all times and upon the whole surface of the globe," [12] scarcely any solution is obtained. Let us rather, with Rütimeyer in his admirable researches on fossil horses,[13] allow our attention to be drawn by these and similar facts "to a close connection between the phases of development in the individual and in the species," that is, to a natural connection.

All who absolutely require a personal God in the current history of creation, draw from these facts no other inference than that their God had the whim of producing at first imperfect and subsequently more and more perfect organisms, and of applying in the development of the last reminiscences of the first.

As worthless as the formula of embryonic types is another, invented by Agassiz, for the shapes in which, in some fossil groups, mechanical and physiological results were imperfectly obtained, and for which provision is made in later organisms by other more adequate and perfect arrangements. These are his "prophetic types." The Pterodactyl is, for example, supposed to stand in this relation towards the bird. Does this quibble aid in the comprehension of either one or the other? Is any rational idea obtained if, besides the prophecy of the Pterodactyl, the geologically antecedent insect is regarded as its prophet, or the bird as the forerunner of the bat? There is no sense at all unless the prophet becomes the progenitor, which in these cases cannot be supposed.

V.

The Standpoint of the Miraculous, and the Investigation of Nature—Creation or Natural Development—Linnæus—Cuvier—Agassiz—Examination of the Idea of Species.

"I hear your message well, it cannot wake my faith.
To faith is miracle her dearest child."*

HAVING quoted these words of Faust, we will proceed without further digression to examine the standpoint occupied by the Natural Philosopher with regard to a domain where the sceptre is wielded, not by the lucid intellect, but by the imagination looking through coloured glasses; not by Logic, but by arbitrary ideas; where the laws of causality are turned upside down; a domain where, indeed, many unquestionably honourable men still feel themselves at home, but which at best fosters only pious self-deception, and indolence of mind.

We must take up a decided position without regard to consequences, as after the discussion of the actual record of the animal world in its three aspects, namely, its present tenantry of complete forms, the evolution of the individuals, and the historical succession during the earlier periods of the earth's formation,—after this superficial work of registration and enrolment, the actual study of our subject must begin.

* "Die Botschaft hör'ich wohl, allein mir fehlt der Glaube.
Das Wunder ist des Glaubens liebstes Kind."

This is, however, the case only with those to whom the miracle of creation is absolutely without existence; whereas an observer, who regards any miracle, however slight, or any sort of disturbance of the order of nature, as possible, must deem his science of Biology complete with the erudition formerly propounded, and subsequently extended by countless items of special information. We cannot therefore do otherwise than give to Goethe's maxim, "Belief is not the beginning, but the end of all knowledge," the interpretation that belief is incompatible with knowledge, and that hence belief in a creation of life is incompatible with the investigation of it.

But if Life did not originate in an incomprehensible manner, it must have been developed. Many decades elapsed before this idea with its consequences could be stated; and in order to comprehend the obstinacy with which the contrary was maintained, and a circle of opinions allowed to take root, against which modern Biology alone has waged a successful war, it is necessary to call to mind some of the chief epochs in the history of Geology, and their representatives. This will naturally lead us to the point whence the shaft of knowledge has been sunk.

After the middle of the last century, Comparative Anatomy, almost independently of systematic Zoology, took a prosperous course, and became far richer in ideas than this descriptive Natural History. One of its maxims, however, was accepted without examination—the constancy and immutability of species; and this maxim forms the centre of the views entertained by Linnæus. The continued authority of this great describer of Nature, is rendered comprehensible only by the confident style as well as by the neatness of his diagnoses, by which, with a single stroke, he put an end to the indefinite character of Natural History, and appeared to contemporaries and posterity as a lawgiver. The exaltation of species as the basis of all systematic comprehension had never been so explicitly proclaimed. His opinions culminate in the maxim,[14] "Reason teaches that at the beginning of things, a pair of each particular species was created." But with Linnæus this said reason looks rather strange, for it is subservient to the strictest Scriptural belief, and he endeavours to harmonize his geological conceptions with this standpoint.

One very effective geological phenomenon was especially striking to him, namely, the upheaval of a great portion of the Scandinavian coast. It proceeds more rapidly than the subsidence of another part; its phenomena are far mightier; and thus the idea might be formed that the continent had risen from the sea in regular progression. "I believe that I am not straying far from the truth," he says, "if I affirm that in the infancy of the world all the mainland was submerged and covered by an enormous ocean, save one single island in this immeasurable sea, on which all animals dwelt and plants grew luxuriantly."[15]

It follows that all species of plants likewise existed in this lovely garden, as it is expressly said that Adam named every animal; consequently all insects must have been assembled in Paradise, but insects cannot be imagined without plants. Linnæus then makes the first attempt at animal geography by making the animals disperse themselves from this centre. But the summary of his idea of species is invariably, "We reckon as many species as the Infinite Being created at the beginning."[16] And his authority was so powerful that the age of Voltaire and of Diderot devoutly accepted this obvious dogma, and transmitted it to posterity as a maxim impossible to question.

Linnæus was, however, so little of an anatomist that in this province Zoology required a completely fresh foundation, and, in the capacity of a second Linnæus, Cuvier stood forth.[17] His school styles itself the school of facts, yet it was by no means without a tincture of philosophy. On the contrary, the definite and simple nature of his principles and deductions could not fail to be imposing. He epitomized the summary of his observations as "Laws of Organization;" and he applied the teleological view, the *principe des causes finales*, with great advantage to the knowledge and restoration of antediluvian animals. The question of the persistency or mutability of species thrust itself forcibly upon him. For this an external cause was given by the Egyptian expedition and the investigation of mummified animals. Etienne Geoffroy St. Hilaire and Lamarck attacked the persistency of species, and held that, especially considering the stability of external conditions, the Egyptian period was far too short for the identity of the mummies with the species now extant, to make it possible to infer the immutability of species; but the question was curtly despatched and silenced by the predominating school of Cuvier.

Meanwhile, Cuvier not only increased the accumulation of facts, but, as we have already hinted, he grouped them so happily and with such philosophical skill that he undoubtedly approached the object at which he aimed—the Natural System. He supplied the first reliable information respecting extinct species. With regard to those which had replaced them in subsequent periods, he was not, as is generally supposed, an unqualified partizan of new creations, but he refrained from any fixed opinion. "I will not," he says,[18] "positively affirm that for the production of the present animals a new creation was required. I merely say they did not live in the same locality, and must have come from elsewhere." Geoffroy Saint Hilaire, on the contrary, does not doubt that the animals now living are descended, by an unbroken succession of generations, from the extinct races of the antediluvian age.

Cuvier's method involved the danger of introducing dogmatism into natural science, and it is therefore justifiable to refer in this place to one of Cuvier's immediate disciples only recently deceased—Louis Agassiz, who in the most rigidly didactic manner adheres to the systematic categories, and invests them with fine-sounding definitions as "embodied creative ideas."[19] According to him, species belong to a particular period in the world's history, and bear definite relations to the physical conditions predominant at the time, as well as to the contemporaneous plants and animals. Species are founded on well-defined relations of individuals to one another and the world in which they live, as well as on the proportions and mutual relations of their parts, and on their ornamentation.

Individuals, as representatives of species, bear the closest relations to one another; they exhibit definite relations also to the surrounding element, and their existence is limited within a definite period. Of genera he says, "Genera are groups of animals most closely connected together, and diverging from one another neither in the form nor in the composition of their structure, but simply in the ultimate structural peculiarities of some of their parts." "Individuals, as representatives of genera, have a definite and specific ultimate structure, identical with that of the representatives of other species."

We may pronounce these definitions to be mere phrases, and inquire with Haeckel: "Of what nature are these 'ultimate structural peculiarities of some of their parts' which are supposed alone to define the genus as such, and to be exclusively characteristic of each genus? We ask every systematizer whether he may not equally well apply this definition to species, varieties, &c., and whether it is not finally the 'ultimate structural peculiarities of some of their parts' which produce the characteristic forms of the species, the variety, &c. In vain do we search in the "Essay on Classification" for a single example of the manner in which, for instance, the genera of oxen or antelopes, the races of hyænas and dogs, or the two great genera of our fresh-water bivalve shells, the Unio and Anodonta, are actually distinguished by "the ultimate structural peculiarities of some of their parts." Several of these definitions given by Agassiz may be interchanged point-blank, so general and merely negative are their statements. He characterizes the classes "by the manner in which the plan of the type is executed as far as ways and means are concerned." The orders, "by the degree of complication of the structure of the types."

These phrases are interchangeable, but, like all dogmatism, they make a great impression on those who from ignorance of the facts are incapable of criticising for themselves, and they are readily quoted to confute an unbelieving investigation of nature by one made in faith.

It might be thought that if the affair were so simple, and systematic ideas so firmly fixed, nothing would be easier than to establish the system. And so Agassiz maintains. He says that if a single species of any of the great animal groups were present, and admitted of investigation, the character of the type, class, family, genus, and species, might be determined. The weakness of this and similar statements may best be demonstrated by examining the basis of all dogmatic system, —the "species." If this idea be mutable, if the species be not given once for all, but variable, according to time and circumstances, the implications of the higher and more general ideas of genus, family, &c., must necessarily ensue. The keenest and most logical criticism on the deeply-rooted scholastic idea of "species" was made by Haeckel,[20] after Darwin, in his classical work on the "Origin of Species," had completely exposed the old doctrine and practice of zoology and botany. In what follows we shall adhere to Haeckel.

We have seen above that Linnæus accepted the Creation as an irrevocable scriptural doctrine, and it is really absurd that many naturalists who have long abandoned any other dogma, should abide by this one. Therefore as the Bible mentions the creation of species, this legend was made the basis of all science. It is true there are not now many who appeal to scriptural testimony.

Those who defend the stability of species rather imagine that, with Cuvier, they are entitled to interpret facts in their own favour; whereas they partly remain unconsciously involved in hereditary prejudice, and partly contrive to be deliberately blind to all that evidently contradicts the immutability of species.

Since Linnæus referred to the Creation, he attributed the individuals to a species, of which the pedigree ascended in direct line to the pair which proceeded from the hand of the Creator. Owing to the state of science in general, an examination of this pedigree was totally impossible in his time; and, indeed, with the strict reliance on sacred tradition, it was scarcely necessary. Cuvier, although a very unprejudiced and cool observer, nevertheless radically accepted the Linnæan definition of species. According to him, the species is the aggregate of individuals descending from one another and from common ancestors, and of those who resemble them as strongly as they resemble one another.[21]

"In this definition," says Haeckel, "to which the majority have ever since more or less closely adhered, two things are obviously required of an individual as belonging to a species: in the first place, a certain degree of resemblance or approximate similarity of character; and secondly, a kindred connection by the bond of a common descent. In the numerous attempts of later authors to complete the definition, the chief stress is laid sometimes on the genealogical consanguinity of all the individuals, sometimes on morphological uniformity in all essential characters. But it may be generally asserted that in the practical application of the idea of species, in the discrimination and nomenclature of the individual species, the latter criterion alone has almost always been employed, while the former has been entirely neglected. Later, it is true, the genealogical idea of the common descent of all individuals of each separate species was supplemented by the physiological definition that all the individuals of every species are capable of producing fertile offspring, by intercrossing, whereas sexual intercourse between individuals of different species produces only sterile offspring or none at all. In practice, however, it was considered quite enough if, among a number of extremely similar animals under investigation, uniformity in all essential characters could be established, and no inquiry was made whether these individuals ascribed to the same species were actually of common origin, and capable, by crossing, of producing fertile offspring. The physiological definition was no more applied in the practical discrimination of animal and vegetal species, than was the pre-supposed common descent from a single ancestral pair. On the other hand, two closely allied forms were distinguished without scruple as two different 'good species,' whenever in a number of similar individuals examined a constant difference could be demonstrated, even though of a merely subordinate character. Here, again, no pains were taken to ascertain whether the two different series were not really descended from common ancestors, and were really capable of generating in conjunction only sterile hybrids, if any."

That this radical condemnation of the post-Linnæan manufacture of species is not too severe, is shown by one fact among others ; that within the fraternity such utter discord as to the limitations of species prevailed, and still prevails, that no agreement can be arrived at respecting the basis of the description of species, the "essential characteristics." Although Agassiz lays down the diagnosis of the species, a decision is required in each case as to the mutual relations of the parts, the ornamentation, &c. As in the absence of birds'-nests, snail-shells, butterflies, &c., it is impossible, when it comes to the erection of species, to pre-determine what may be the "essential characteristics" of the species they are to form, subjective opinions and arbitrary decisions have full play; and within a certain domain, well known by its forms, there are among the systematizers no two authorities who are agreed as to the number of species into which the material before them should be divided.

The most unbridled license in the manufacture of species prevailed, however, among the Palæontologists during a period when, in the endeavour to fix the subdivisions of geological strata as accurately as possible by means of their organic contents, the separation of species was carried incredibly far, into the most minute and often into individual deviations. A certain mutability of species could not fail to obtrude itself on the most purblind eye; ramifications were made of sub-species, sports of nature, and varieties characterized by "less essential" peculiarities acquired by means of climate and inheritance. There was, however, always a reservation that their crosses with one another and with the main species should produce fertile offspring, whereas towards other species their relations were identical with those of the main species. Of course, in this separation of the species into sub-species, subjective opinion was even less fettered by tradition and law than in the definition of species. The literature of ornithology during the last forty years could furnish thousands of the strangest examples of the Babel-like confusion which was thus introduced.

There is no question that a great, perhaps the greater, number of organisms now existing are in a condition in which, according to their internal and external relations, they may be characterized by Natural History as so-called species, and for the purpose of recognition and scientific treatment in general, must needs be so characterized. But this stability, as may be shown both directly and by analogy, is under all circumstances only temporary, and we have whole classes of organisms to which it is impossible, even with the widest reservations, to apply the old idea of species, with its immutability of essential characteristics. If we are able to furnish incontrovertible proofs of the existence of such non-specific groups, the old system and the dogma of species are once for all set aside, and the positive basis of a new doctrine is secured. This evidence is supplied in two directions. Some classes of organisms in their present state vacillate and fluctuate in form, in such a manner that it is utterly impossible to fix the characteristics of species or genus. They are in an extreme grade of mutability, which, in others, has given way to an apparent state of repose. Other series of facts, exhibiting the most obvious mutability of species, are displayed by certain antediluvian groups in the succession of forms called "species."

Even before the appearance of Darwin's work on the "Origin of Species," Carpenter, in the course of his researches on the Foraminifera, arrived at the conclusion, proved in special instances, that in this group of low organisms which secrete the most delicate calcareous shells, there could be no question of "species," but only of "series of forms." Forms which the systematists had reduced to different genera and families, he beheld developing themselves from one another. These Foraminifera are, however, so simple in structure, the history of their individual evolution or Ontogenesis is, as yet, so little known; they contribute so little microscopic detail, which might formulate the law of transmutation of species, that the champions of persistency of species might still seek refuge in the assertion that Carpenter's series of forms are mere varieties, and only prove that the true "species" have not yet been found.

We may now turn with advantage to the class of the Spongiadæ, the importance of which in the question of species I was the first to point out.[22] With them, as I summed up my researches, it is not as with the Foraminifera, merely an affair of the general habit of the form, of the variable grouping of the chamber systems; but the variability exists still more specially in the microscopic detail than in the coarser constituents. In the Foraminifera we may speak of microscopic forms, but not properly of microscopic constituents. But in the sponges we discern the transformation of the finer morphological constituents, the rudimentary organs, and we thereby gain an insight into the mutability of the whole. In this respect the calcareous sponges are somewhat differently circumstanced from the rest, and from the silicious sponges in particular. In the former, the variability of the microscopic parts is limited to a smaller circle of forms, whereas the habit of the series of individuals is incredibly pliable. This pliability of the whole body is not lacking in the silicious sponges; in the genus Tedania, for instance, established by Gray from some of my earlier Reniera, we see how their stubbornly coherent needle-like forms recur from Trieste to Florida and Iceland, under the most heterogeneous disguises. In some varieties, however, one of these spicula already manifests a tendency to deviations.

This very point, the possibility of tracing in detail the metamorphoses of organs, which, on the assumption of their stability, appeared to provide the system with the most substantial basis for the erection of genera and species, renders the investigation peculiarly attractive. Even among the Algierian sponges, I have adduced striking examples, and they accumulate in proportion as the horizon is extended. We arrive gradually at the conviction that no reasonable dependence can be placed on any "characteristic;" that with a certain constancy in microscopic constituents, the outward bodily form, with its coarser distinctive marks, varies far beyond the limits of the so-called species and genera; and that, with like external habits, the internal particles, which we looked upon as specific, are transformed into others, as it were, under our hands. "Any one"—thus concludes this section of my work on the Fauna of the Atlantic Sponges,—"who, with regard to sponges, makes his chief business the manufacture of species and genera, is reduced *ad absurdum*, as Haeckel has shown with exquisite irony in his Prodrome to the Monograph on the Calcareous Sponges."

In my specific researches I confined myself essentially

to the silicious sponges, and by thousands of microscopic observations, by measurements, by drawings, by facts and inferences, had produced evidence, which acute opponents of the immutability of species had not brought forward before me, that in these sponges, species and genera, and consequently fixed systematic unities in general, had no existence. The other division of the same class, the calcareous sponges, had been treated with unrivalled mastery by Haeckel in his Monograph.[23]

He was able not only to confirm my statements, but, owing to the smaller compass and the greater facility of observing the group selected for study, to advance with more sequence and continuity from the observation of details to the whole, to portray its morphology, physiology, and evolutionary history with the utmost completeness. He then challenged the obstructive party with the assertion that, according to subjective opinion, either one or 591 species of calcareous sponges might be accepted, but "that no absolute species exists, and that species and varieties cannot be sharply separated." Whoever after these demonstrations cleaves to the phantom of species, without either proving that the facts have been falsely observed, or that they may be interpreted otherwise than in favour of the stability of species,—whoever, as Agassiz has recently done, ignoring any such researches, publicly asseverates that in no single case has the mutability of any species been exhibited,—scarcely preserves the right to participate in the great controversy by which Natural Science is now perturbed.

There is, however, as we have already mentioned, a second direction in which the mobility of "species" must be demonstrated, not the direction of breadth, but of height and depth. This mutability of the Spongiadæ affords the extremely important evidence that, so to speak, an entire class has, even now, not attained a state of comparative repose. But to confirm the mutability of species, evidence of mutability in lapse of time is justly demanded; the transition of the forms succeeding one another historically in the strata of the earth.

A highly instructive example of the transmutation of species occurring in the lapse of time, and one which may at all events be banished from the limits of varieties, is offered by the Tellina (Planorbis multiformis) occurring in the fresh-water chalk of Steinheim, in Würtemberg. The deposit, derived from the Tertiary period, contains the residue of a small lake, and may be divided into about 40 petrographically distinguishable layers. "In the whole series of strata," says Hilgendorf,[24] "the varieties of Planorbis multiformis are distributed in such a manner that individual layers are characterized as successive strata, by the exclusive occurrence or by the predominance of single or several varieties which, within the layer, remain constant or slightly variable, but towards the limits of the next layer, lead by transitions to the succeeding forms. The intermediate layers furnish evidence that the other forms originated by gradual metamorphosis from the earlier ones; they moreover render it possible to range form to form, and to trace the evolution backwards; hence it becomes manifest that what above seemed distinctly divided, meets below. Thus arises a pedigree richly endowed with main and side branches.". The forms diverge so greatly, and are so constant in the main zones, which tell of periods of repose, that, in accordance with the old conchological practice, they would be unreservedly claimed as species, if the connecting links were not too conspicuous and the territory too circumscribed, and if the geological period, which must, however, be reckoned at least by thousands of years, were not considered too insignificant.

But what the case of Steinheim exhibits in miniature was taking place on a large scale during the great geological periods, and the zeal of some Palæontologists, such as Waagen, Zittel, Neumayr, Würtenberger,[25] has had the effect of proving, at least with respect to the important division of the Ammonites, the utter impossibility of separating them into "species." The study of Ammonites has shown that from fossil remains important inferences may be made as to the whole organization, and that, in combination with the observable modifications of the shell, simultaneous and profound metamorphoses of certain determining soft parts must have taken place. If it is now proved, as it has been by these investigators, that the so-called "species" which characterize the great Jurassic and cretaceous formations, are connected in the same manner as the varieties of the Steinheim snail, as mere morphological series of variable constancy and duration, they who will not allow even this evidence to rouse them from their innate drowsiness, are like the ostrich which prefers not to see the danger. Neumayr is such a cool and cautious observer, that he allows nothing to pass current but that which is absolutely certain. It is true he holds it to be "extraordinarily probable" that in all forms these gradual transitions have taken place, yet in one case only does he demand unqualified assent, namely, that he has proved " that Perisphinctes aurigerus (Opp.) of the Bathonians, and Perisphinctes curvirostrus of the zone of the Cosmoceras Jason (Rein), are connected in such a manner by intermediate occurrences that it is impossible to draw a limit.

L. Würtenberger applied his researches to thousands of samples from the groups of the Planulate Ammonites with ribbed shells, and of the Armate Ammonites with prickly shells. In summing up his results he says, among other things: "How among the Ammonites of the Planulate and Armate groups, the species are to be branched off from one another, I should be reluctant and unable to give any instructions, for to me this question appears utterly hopeless. For in groups of fossil organisms, in which, as in the present case, so many connecting links between the most extreme forms are actually before us, that the transition is regularly carried on, the species is far less susceptible of apprehension than in the organic forms of the present world, which at least denote the existing limits of the great pedigree of the organic world. With respect to these fossil forms, it is fundamentally indifferent whether a very short, or a somewhat longer portion of any branch be honoured by a special name, and looked upon as a species. The prickly Ammonites, classified under the name of Armata, are so intrinsically connected, that it becomes an impossibility to separate the accepted species sharply from one another. The same observation applies also to the group of which the manifold forms are distinguished by their ribbed shells, and termed Planulata." It has further transpired that the Armata, or Custata, originated from the Planulata.

We shall return later to Würtenberger's preliminary communications. It was our object here to inform our readers how and where modern natural inquiry sets aside the phantom of species, and to enable them to judge for themselves what series of observations are opposed to the asseverations that in no single case has evidence been given of the transition of one species into another. For the old school falls into the dilemma of proclaiming whole orders and classes to be "species," and the species, formerly so beautifully defined, to be varieties.

The untenableness of the physiological part of the definition of species has been conclusively shown first by Darwin and afterwards by Haeckel. It is known that even in a state of freedom good species not infrequently breed together, and that domesticated species, such as the horse and the ass, have been crossed for thousands of years. But hybrids, the produce of this intercourse, were supposed to be only exceptionally fertile, and at any rate not to produce fertile progeny for more than a few generations. On the other hand, it was considered certain that the produce of crosses among varieties are fertile in unbroken succession. The dogma of the sterility of hybrids was formed without any experimental or general observation, and by ill-luck was apparently confirmed by the most ancient and best known hybridizations of the mule and the hinny. To this familiar example, in which the fertility of hybrids proves abortive, we will oppose only one case of propagation successfully accomplished in recent times through many generations; that, namely, of hares and rabbits, two "good species" never yet regarded as mere varieties.

The numerous and varied forms of the domestic dog

were pronounced *ex cathedrâ* to be varieties of the same species, as their crosses are productive. But after reading Darwin's careful comparison of the reports as to the relations of certain species of wolves with the dogs of savage nations, and of the European wolf with the Hungarian dog, we must agree with Darwin in thinking it as extremely probable that in various parts of the world, and at various periods, wild species of the genus Canis were domesticated, of which the crosses produce fertile progeny to an extent almost unlimited.

It is the same with the domestic cat. With the forms of the European domestic cat, the case is such that it is scarcely possible to doubt its origin partly from a Nubian species, and partly from the European wild-cat. The inferences thus moved in a circle; forms belong to the same species, because they may be fruitfully crossed; and because they may be fruitfully crossed, they belong to the same species; and, on the other hand, because such and such forms, when crossed, produce no fertile progeny, they constitute different species; and because they are different species, they generate no fertile offspring. The cases of persistent fertility in hybrids are certainly not frequent, but they are nevertheless so well certified that the contrary statement is in plain contradiction to the facts. But conversely, the proposition that mongrels, the products of crosses among varieties, are fertile, thus generally stated, is likewise untenable. The variety which has been evolved in Paraguay from our domestic cat, pairs no longer with its ancestral stock, nor does the tame European guinea-pig with the wild ancestral stock of Brazil.

But even if, in general, crosses between varieties are

more easily effected, and more often produce fertile offspring than the unquestionably rarer crosses of species the frequent failure of crosses between species completely accords with the modification of species in the lapse of time, as shown above. Provisionally, let us hold nothing to be established but that, as to fertility and the capability of persistent reproduction, the conditions of mongrels and of hybrids are essentially similar and differ only in degree, and that on these properties, no closer definition or limitation can be founded.

If the older definitions of species go back to Paradise, and derive existent species lineally from ancestral progenitors, miraculously created from the first and never modified, the ingenuous statements of Linnæus show that all this was accepted as self-evident, and that no thought was given to the proof, which would indeed have been impossible to obtain. A letter from George Forster to Peter Camper, dated May 7th, 1787, proves however that, even in the last century, the voices of more far-sighted naturalists were raised against this superficial treatment of the idea of species. Systems, he said, were founded on this idea, yet everything was uncertain as long as this expression was not irremovably fixed. But hitherto all definitions of this word were hypothetical, and in themselves anything but clear. If we are to accept as many species as were created, how is a created species to be distinguished from one produced by the intermixture of several others? To fall back upon the Creation is to lose oneself in the Infinite and the Impalpable. "This will never enable us to understand anything; and definitions which rest on an inexplicable foundation, on a mystery, ought to be proscribed from science for evermore."

Without owning allegiance to any theory whatever, we are constrained to recognize the fact, that in various groups of organisms there even now exists such an instability of form, and such a degree of variability, that it is patent how constrained and artificial is their systematic separation. In many other groups, in most orders of the Mammalia, for example, this phase of mobility has been replaced by a certain quiescence, and the forms now presenting themselves for observation and comparison are so well defined from one another, that they fit into the system without difficulty as "good species." But if the "good species" are to be judged by the experiences made in regard to the "bad" ones, and if the preposterous hypothesis is not laid hold of, in contravention to all healthy human understanding, that "good species" originated in a miraculous manner inaccessible to our cognition, whereas the "bad species" are susceptible of analysis,—the other alternative alone is possible, that, as Haeckel says, if we knew them thoroughly, all species without exception would, in the sense of the species-makers, be "bad species." We are also acquainted with a sufficient number of bad species to be capable of inferring the general law with certainty. Nevertheless, all further corroboration and discovery of bad species is acceptable. Regarded formerly by the systematists only as incumbrances and as stones rejected by the builders, they have now become the corner-stones of science.

Is species therefore, we again inquire, to be entirely abandoned? Not so, for several reasons. Even assuming that so-called good species, in the sense of the systematists, have no existence, human intellect, in the endeavour to obtain a general view, would be compelled to denominate the forms, unless all scientific treatment was to be rendered impracticable. But the retention of species is moreover scientifically justifiable and necessary, if only the determining impulses be taken into account, and the definition reduced to harmony with reality. Species is not constituted merely of analogous individuals, for even the sexes, in the course of development, and without transformation, diverge considerably from one another.

But if we remember the transmutation of shape taking place by stages in organisms subject to metamorphosis, and the regular sequence of forms alternating with one another in heterogenesis, we shall be obliged to speak, not of individuals, but of the cycles of reproduction which comprise the various phases and series of individuals. These remain persistent as long as they exist under the same external conditions. How far time in itself affects existence and decay is unknown. At any rate, time, as well as the external conditions of time, is a factor in the mutation of species. While we regard species as absolutely mutable, and only relatively stable, we will term it, with Haeckel, "the sum of all cycles of reproduction which, under similar conditions of existence, exhibit similar forms."

VI.

Natural Philosophy—Goethe—Predestined Transformation according to Richard Owen—Lamark.

WE have hitherto confined ourselves essentially to the contemplation of the phenomena of the animal world as facts, avoiding as far as possible any examination of the correlation of these facts, or any criticism of the attempts to explain them. It was nevertheless necessary to single out from the history of our science some few impulses of which the after-effects extend to the present time, and of which a knowledge is conducive to the comprehension of prevailing views, tendencies, and prejudices. For this reason we again revert to the evolutionary history of Biology and Comparative Anatomy, that we may trace the present currents to their sources. Since the middle of last century, there has been no lack of leading ideas in the organic natural sciences, such, for instance, as are contained in Buffon's magnificent project of a picture of the world. But if it is a question of a single comprehensive solution of the organic world, we are at once reminded of the claims preferred by Natural Philosophy in the first decades of this century, to explain the universe; to derive from the whole, not only matter in the abstract, but the being and origin of organic bodies. When the Philosophy of Identity began to found the laws of the Mind without the study of the body, and in its own fashion had proved the identity of the corporal and spiritual world by means of imponderables and non-organic bodies, their constructions necessarily extended to organisms.

This attempt to generalize the principles of Schelling was made by Oken [27] when in his system he conceives all Nature to be a process of evolution. In his opinion, natural science is the science of the eternal modification of God, that is of Mind, in the world, and is thus in the widest sense, Cosmogony. Everything, when contemplated as part of the genetic process of the whole, involves, besides the idea of existence, also that of non-existence, or position and negation, as it rises into a higher idea. These contrasts include the category of polarity, which manifests itself in motion, the life of all things. The simpler elementary bodies aggregate into higher forms, which are mere higher powers of the former, as their causes. Hence the various classes of bodies represent parallel series, each corresponding with and modifying the order of the other; classes of which the rational arrangement follows with inherent necessity from their genetic coherence. But in individuals, these lower series again become apparent during the period of development. The antagonisms in the solar system of the planets and the sun, repeat themselves in plants and animals; and as light is the principle of motion, the animal has the advantage of independent motion, above the vegetal organism which pre-eminently belongs to the earth. Embryology receives its due in a general proposition. "Animals perfect themselves gradually, adding organ to organ in the self-same manner as the individual animal is perfected." But in Man, as the highest animal, the whole animal world is contained; he is the actual Microcosm.

If Natural Philosophy be the expression and logical connection of all well-observed facts, we could not now designate as Natural Philosophy, Oken's well-rounded system, laid down in 3562 propositions, with their inferential conceits of Position, Negation, and Polarity, the absolutely meaningless formula of + 0 — without any real penetration of the subject-matter. Various and important incitements to research were nevertheless supplied by it, and we have been the more anxious to call attention to this system, as it implies at least as much as the vague formulæ and ideas of "intrinsic development," the "principle of progress," the "conversion of the lower into the higher," and the whole litany of indecision and indistinctness.

In this chapter we shall not adhere to chronological succession, but merely characterize various theories of organic nature; and we may therefore now revert to Goethe, who in Haeckel's opinion forestalled his age on the great question which forms the subject of this book, and deserves to be honoured as the independent founder of the theory of descent in Germany.[28] We cannot ascribe this importance to Goethe, for we must deny the very cardinal-point on which Haeckel lays most weight,—that Goethe regards species not merely as modified phenomena of the variable idea of the genus, but as the sum of bodies modifiable in the concrete. What principally induces us to make detailed mention of Goethe is his penetration of the idea of type, which since the time of Buffon had been for two generations the lodestar of a higher research unknown to the pure systematizers. Goethe elaborated this idea in his own mind on the basis of a certainly remarkable special knowledge of organic matter, and undeniably reached the threshold of the solution. That his scientific activity was a necessary effusion of his nature, I have demonstrated in the treatises here cited. Additional evidence has been given by Helmholtz and Virchow.

Goethe's notes on his position towards nature, and his researches, comprise a period of more than fifty years. About the year 1780, there appears, under the title of "Die Natur," a sort of Hymn to Nature, concluding with the beautiful words which make him seem a pure Pantheist: "She placed me in it; she will also lead me forth; I trust myself to her. She may dispose of me. She will not hate her work. I spake not of her. No, whatever is true and whatever is false, she spake it all. All is her fault, and all is her merit." And shortly before his death, in March, 1832, he threw his whole soul into the scientific controversy as to the different methods of the investigation of nature and the fundamental principles of study, which rose high in the midst of the French Academy between the two renowned representatives of the inductive and deductive tendencies, Cuvier and Geoffroy St. Hilaire. What Goethe here laid down in the evening of his days, is a sort of scientific profession of faith, and it inspires the greatest admiration to behold the venerable octogenarian standing on the pinnacle of time, and above all parties, with the same principles which with his own powers he had framed for himself five-and-forty years before, in the prime of manhood.

In the height of his genius, when Goethe, standing at the centre of the life of Weimar, frequently withdrew from the bustle of the town and court, he received the first suggestions of the "Metamorphosis of Plants." He was irresistibly attracted to the varying phenomena of vegetal life, and he could but muse on the implied unity and rule underlying this variation. This was a fresh source of agitation, which pursued him when, in 1787, he forcibly tore himself from the influences of Weimar and fled to Italy. There, in Sicily, he found the solution of the riddle: the leaf seemed to be the rudimentary organ of vegetal structure. And when, after his return, a new star rose for him in Christiana Vulpius, he laid down the quintessence of his ideas on the Metamorphosis of Plants in that exquisite poem, of which the lines—

"All forms have a resemblance, none is the same as another,
And their chorus complete points to a mystical law,
Points to a sacred riddle,—" *

are present to all who ever made themselves acquainted with the muse of Goethe. He now saw in the various parts of the plant what he had learnt to see with the eye of the imagination, which he considers essential to the Naturalist,—the harmonizing principle. "The same organ may be expanded into a compound leaf, or contracted into a simple stipule or scale. According to different circumstances, the self-same organ may be developed into a peduncle or an unfruitful branch. The calyx, by over-hastening itself, may become the corolla, and conversely, the corolla may approximate to the

* Alle Gestalten sind ähnlich, und keine gleichet der andern,
Und so deutet der Chor auf ein geheimes Gesetz,
Auf ein heiliges Räthsel——

calyx. Thus the most varied structures of plants are rendered possible, and he who in his observations keeps these laws always before his eyes will derive from them great alleviation and advantage." These few lines contain the pith of the doctrine of the Metamorphosis of Plants which so greatly agitated his contemporaries during the first quarter of this century. The many-sidedness of the idea made it inevitable that the notion, once grasped, should extend to the remainder of the organic world. Before Goethe, no naturalist had regarded insects otherwise than as a given sum of individual forms, distinguishable by certain definite characteristics. Their internal structure had certainly been disclosed by some few great men, such as Malpighi, Swammerdam and Lyonet, but a real comparison of species and genera had never been contemplated; still less an explanation of the body by its parts. This Goethe accomplished, and with true genius; for to his theory, and with perfect truth, the rings which in the insect are ranged from the head to the tail, presented themselves, like the vegetal organs, as mere modifications of one and the same rudimentary organ. There, the leaf in the abstract, the primordial leaf or plant—here the ring.

With this—it was in 1796, in the discourses on the project of a general introduction to Comparative Anatomy—he enunciated a truth which was not recognized till more than forty years later, by one of the most distinguished zoologists, Milne Edwards, and applied to the knowledge of the animal world. This is the idea of the development of organic beings by the heterogeneous evolution of their fundamentally similar parts. Of this the caterpillar and butterfly serve as an example. "Imperfect and evanescent a creature though the butterfly may be as to its species, when compared to the mammal, in the metamorphosis which it accomplishes before our eyes, it nevertheless exhibits the superiority of a more perfect over a less perfect animal. This consists in the decisiveness of its parts, the security that none can be put or taken for the other; that each is destined for its function, and remains constant to it for ever." Now, however, in the most perfect creatures, the Vertebrata, there appeared before Goethe's eye, a similar rudimentary organ, metamorphosing itself within the individual; this was the vertebra. He followed it in its transformations along the vertebral column. Impossible as it may be, by placing together the first vertebra of the neck with the last tail bone to infer their identity, it becomes manifest in the gradual transition.

But what lies in front of the first vertebra of the neck? Is the cranium something absolutely different, something new, not identical with the vertebral column? This was another perturbing thought which pursued Goethe's every footstep. He pondered and compared; it could not be otherwise; the cranium must belong to the vertebral column, must be nothing more than a part of the vertebral column. Through the vacillations of his conceptions, he was, as he later expresses himself on another occasion, "as an honest observer transported into a sort of frenzy." Then, when in 1790 he picked up a bleached sheep's skull in the Jewish cemetery at Venice, "the derivation of the cranium from the vertebral bones was revealed to him." The more special history of Comparative Anatomy has shown how extremely fruitful was this supposed discovery, although the subject is far more complex than Goethe and his followers imagined.

We must commemorate yet another genuine discovery made by Goethe, which exhibits his very peculiar method. It relates to the inter-maxillary bone in man. About 1780, he was studying osteology at Jena, under the guidance of Loder, an anatomist of some renown. It is evident that all higher animals possess a bone, the so-called inter-maxillary bone, supporting the upper incisor teeth. "The strange case now occurred," relates Goethe, "that the distinction between apes and men was made by ascribing an inter-maxillary bone to the former, and none to the latter; but as this part is mainly remarkable as the upper incisor teeth are set in it, it was inconceivable how man should have the incisor teeth and lack the bone." It was inconceivable to him because, from the comparisons of Nature, he had framed the idea "that all divisions of the creature, singly and collectively, may be found in all animals." To make man an exception, not to be measured by the same pattern, was repugnant to his mind. Man must have an inter-maxillary bone; and, contrary to the opinions of the greatest anatomists of that period, such as Peter Camper, he demonstrated how in man this inter-maxillary bone, although it subsequently becomes almost undistinguishably anchylosed with the actual supra-maxillary bone, nevertheless exists, quite distinctly, as a separate part during development and early infancy.

From this narrative we have gained a good deal. In the contemplation of individuals and details, Goethe found no pleasure. Nature and natural objects, as existent and complete, merely inspired the wish forthwith to examine their origin and its cause. To judge of things by their final causes, according to an assumed purpose pre-determined by Providence, he deemed "a melancholy expedient" which must be entirely set aside. For this method of contemplating Nature, as pursued by him, in which all living things are to be conceived as intrinsically connected, the external as an indication of the internal form, he created the name of Morphology, the doctrine of form. He examined "how Nature lives by creating;" and from amazement at the eternal formation and transformation, from the perplexity into which he was plunged by the manifold variety of forms, we see him emerge by seeking and finding primordial forms.

Even before the realization of the metamorphoses of plants, we find him surrounded by bones and complete skeletons in his scientific ossuary at Jena; he thought he had found a lodestar in the erection of an anatomical *Type*, an universal symbol, "in which the forms of all (vertebrate) animals were potentially contained, and by which each animal may be described according to a certain arrangement." "Experience must first teach us which are the parts common to all animals, and wherein these parts differ. The idea must control the whole, and in a genetic manner deduce the universal model." Thus by an abstract of the individual, we are to possess ourselves of a certain archetype. As man could not be taken as a standard for animals, and conversely, the infinite complexity of man could not be fully explained by animal organization, something fluctuating between the two must be summoned to solve the problem. To this archetype, itself incapable of representation,—to this abstraction, and to this alone,—Nature, according to Goethe, was bound to adhere in her work of creation, "without being able, in the slightest measure, to break through or overleap the circle."

If it be attempted to make it appear that Goethe actually proclaimed the doctrine of Descent, or was even in a poetical sense its inspired prophet, either too much value is attributed to his enunciations of "ceaseless progressive transformation," and such like, or the sense which he connected with them is not appreciated. Now let us take the following passage, which Haeckel looks upon as decisive. "Thus much we should have gained ; that we may fearlessly affirm all the more perfect organic beings, among which we include Fishes, Amphibians, Birds, Mammals (and at the head of the latter, Man), to be formed according to an archetype, which merely fluctuates more or less in its very persistent parts, and moreover, day by day, completes and transforms itself by means of reproduction." Is it here meant, perchance, that the persistent are contrasted with the non-persistent parts ? By no means.

Even prior to Geoffroy Saint Hilaire, Goethe had spoken of a law, which is, however, no law, nor even an expression of facts, namely, that Nature in her work has to deal with a given quantity of material to which she must adapt it. He does not seem to have been aware that Aristotle had affirmed the same, that Nature, if she enlarged an organ, did so only at the expense of another. A second of the supposed fundamental laws discovered by the Frenchman, that an organ would sooner perish than resign its place, was likewise instituted by him at the same time.

Thus, in Goethe's opinion, nature always makes use of the same parts. Nature is inexhaustible in the modification and realization of the archetype; but to that which has once attained realization cleaves the tenacious power of persistency, a *vis centripeta*, of which the profound basis is beyond the influence of anything external. Hence, if he speaks of daily completion and transformation by means of reproduction, he understands, with respect to the animal which has attained realization, merely that course of development or metamorphosis which is an image of inexhaustible phenomenal nature. The influences which Nature has exercised upon the parts, he pictures to himself as still present; but of an actual transformation of existing species into new ones, such as is required by the modern Darwinian doctrine of Descent, Goethe does not speak at all.

In his view, what was it, then, that was to be transformed? Surely not the archetype. He says, indeed, "Thus the eagle fashioned itself by the air for the air, by the mountain top for the mountain top. The mole fashions itself to the loose soil, the seal to the water, the bat to the air;" and generally, "the animal is fashioned by circumstances to circumstances." But the illustrations which he gives in the Sketch of A.D. 1796, show plainly that he thought, not of any transformation of existing forms, but of mere modes of manifestation of the type and archetype as they exist in given species. He then says, "The serpent stands high in organization. It has a decided head, with a perfect auxiliary organ,—a consolidated lower jawbone. Only its body is indefinitely long; and the cause of its being so is that it expends neither material nor power upon auxiliary organs. As soon as these make their appearance in another form, as, for instance, in the lizard, though only short arms and legs are produced, the indefinite length must at once contract, and a shorter body takes its place. The long legs of the frog necessitate a very short form for the body of this creature, and by the same law, the unshapely toad is laterally extended." It is well to bear in mind this somewhat trivial passage, that we may not see more in the poetic glorification of the Metamorphosis of Animals than it really contains.

When Goethe says in the magnificent poem:

> "Hence, each form conditions the life and acts of the creature,
> And each fashion of life, with reflex forcible action,
> Works on the form:" *

it sounds, as we must admit, extremely seductive. But we are sobered, or rather led to the right standpoint, by reading his fascinating remarks on d'Alton's skeletons of the rodents (1824). It is there made manifest that Goethe had not the remotest idea of an actual transformation of a rodent into any other animal by the force of external influences.

The reader may judge for himself. "Let us contemplate the animal in the neighbourhood of water; as the so-called water-hog it wallows, pig-like, on the marshy shore; as a beaver it is seen building by fresh waters;

* Also bestimmt die Gestalt die Lebensweise des Thieres,
Und die Weise des Lebens, sie wirkt auf alle Gestalten,
Mächtig zurück——

next, still requiring some degree of moisture, it burrows in the earth, and at least loves concealment, hiding with coquettish timidity from man and other animals. Finally, when the creature arrives at the surface, it hops and frisks, so that it carries on its existence erect, and even moves to and fro on two feet with marvellous rapidity. Transferred to completely dry land, we at last find the decisive influence of the airy eminence and the all-vivifying light. The animal is endowed with the greatest ease of movement; it acts and works with consummate skill, until a bird-like motion passes into an apparent flight."

Thus does Goethe elaborate the influence of environment and external conditions upon the modifications of form; it is in vain to look for the actual forms that are modified. The beaver is not transformed into the mouse-like burrower, the mouse into the jumping mouse, nor the jumping mouse into the squirrel, nor does the latter become a jerboa; but the "ceaseless progressive transformation" is perceptible only to the eye of the imagination. In reality, moreover, Goethe sees only adaptation. Greatly as he is inclined to attribute modifications to the effect of external conditions, he speaks with no less decision on the contrary side. "The parts of the animal, their relative form, their conditions, their special characters, determine the requirements of the creatures' existence;" and if within the restricted circle of forms, we nevertheless find that infinite modifications of form become possible (Sketch, 1796), this is only to be deduced from the individual species exhibited as modifications of the archetype, by Nature, ever one and ever creative.

With the word Species, we reach the most important point in our account of Goethe's theory of nature; if indeed we have not already unquestionably proved that he can in no way be regarded as a true precursor of Darwin. Darwin and his adherents maintain the variability of the so-called vegetal and animal species. The question is simply whether Goethe was or was not, like his contemporary Lamarck, convinced of this mutability. If he says on one occasion that "from the seed, plants are developed, ever diverging and variously determining the mutual relations of their parts," this is ambiguous in itself; it may refer either to the origin of new species, or to the variability of species by nature immutable. Another time he speaks of the "purpose of Nature" in the horse.

I can find but one single passage in Goethe's writings in which there is a question of an actual transformation of a creature, if not into a new species, at least into a very marked and persistent variety. In 1820, a Dr. Körte gave a description of a primæval bull found in the neighbourhood of Halberstadt, and instituted comparisons and reflections, how under the influence of domestication our highly modified cattle had been evolved from the former. This relic, and another in Thuringia (1821), which latter specimen was obtained by him for the Museum at Jena, gave him an opportunity of coinciding with Körte, and of illustrating by an actual incident, the possibility of this doubtless easy transformation.

But from this to the transformation of species there is still a long way, and Goethe did not traverse it. We have just seen that the idea of deriving single animals now existing from extinct "ancestral races," was not unfamiliar to him. Nor would his remark, "For we have the most distinct remains of organic creatures which were unable to perpetuate themselves by active reproduction," exclude his having accepted generally the immediate connection, based on direct reproduction, of the animal world with fossil races entirely differing in structure. For it is quite true that many species, genera and groups, passed through, not their prime only, but also their decline and total extinction antecedent to the present era.

Yet more. In "Aphoristic Annotations," which he terms problems, written previous to the year 1823, he speaks of "characterless races, which it is scarcely permissible to assign to a species, as they lose themselves in boundless varieties," and he contrasts them "with races possessed of a character, which they exhibit afresh in all their species, so that they may be ascertained in a rational method." Goethe rests on this fact to illustrate his idea of metamorphosis; and we have no right to explain the characterless or "disorderly" races in a Darwinian sense, as being those of which the forms are not established, while those which possess a character are divided into easily distinguishable species, because a host of intermediate forms have succumbed in the struggle for existence. He gave this problem to his intelligent young friend, Ernst Mayer, that he might work it out, and impart his reflections to his instructor.

Mayer says: "The more readily the former (the genera possessing character) are arranged, the more difficult it is to dispose of the latter (those which possess no character). But any one who observes them with earnestness and persevering zeal, and is not totally deficient in intuitive tact, cultivated by exercise, far from being perplexed by them, will assuredly, amidst all their varieties of form, very soon detect the true species and their characters. But if indeed, in any one genus rich in forms, no limit to which nature herself adheres should be discovered, what should hinder us from treating it as a single species, and all its forms as so many varieties? As long as the evidence is wanting, which it is not likely will ever be produced, that no species whatever exists in nature, but that every, even the remotest form, may be evolved from the other by intermediate links,—till then we must be allowed to rely upon the course already indicated. Let the master now instruct the scholar, or, according to ancient custom, support him." And he does support him, for in his morphological writings he adopts his pupil's enunciations on the problem as a testimony of entire community of mind and soul.

There can be no question that Goethe's thoughts on organic nature were more profound than those of his contemporaries. But we must not forget that the cardinal idea of a modifiable archetype prevailed among eminent men both before and with Goethe, as I have shown in my little work known to the profession, "The Development of Comparative Anatomy" (Die Entwickelung der vergleichenden Anatomie, 1855). If in his popular lectures, Peter Camper amused his audience by a diagram in which he evolved a beautiful female figure from a horse; if he says that he is so entirely absorbed in studying the whale and comparing it with the human structure that every girl, pretty or ugly, appeared to him like a dolphin or a cachelot; this was because he started from an archetype or fundamental form. Goethe was only more consistent, and notwithstanding the "painful consequences," insisted on the inter-maxillary bone in man as in the ape.

Goethe says in 1807: "If plants and animals be contemplated in their most imperfect condition, it is scarcely possible to distinguish them. This much, however, we may say, that from a kindred so close as scarcely to be discriminated, emerge creatures which, as plants and animals, are perfected in two different directions, so that the plant finally attains its glory in the tree, durable and rigid; the animal in man, in extreme mobility and freedom." But this is nothing more than the repetition, symbolically embellished, in Goethe's "method of investigating, knowing, and enjoying," of a proposition already propounded by Buffon fifty years before, and subsequently varied in many ways.

Nor is it Goethe who, in his Sketch of 1796, first urges the very suggestive comparison of identical organs in the same body; this was already done by Vicq d'Azyr in 1786. In a word, the conception of type, archetype, ground-plan (*dessein primitif*), was an acquisition of that age, which was merely expressed by Goethe in a more pregnant and many-sided manner, and which appears the more alluring as he combined with it the idea of motion and mobility, but, owing to his excessive craving for symbols, he did this in a figurative sense.

When Goethe imagines that he has discovered "laws," he labours under the same delusion as that in which naturalists have rocked themselves from the last century down to the most recent times, when they accept a mere corroboration of facts for an explanation of those facts, for their reduction to their causes. He is acquainted with a "spiral tendency" and a "vertical tendency" in plants, and they at once become "fundamental laws of life." Now in root and stem we undoubtedly see a vertical tendency downwards and upwards; we see convolutions and tendrils; we have, moreover, been able to analyze these facts into simpler physical and physiological phenomena, without having arrived at the innermost cause, the actual law.

Goethe's opinion as to man's place in Nature is implied in what has been already said. That he, a creature and a product of Nature should form an exception to the animal so obviously resembling him, he could not admit. He must remain therefore unconditionally within the type, "of which the parts are perpetually modified in all races and species of animals." But we have now, I think, furnished sufficient evidence that this and similar enunciations apply only to the potential variability of the archetype which has found expression in the races and species. Hence man also is to him a product allied to the animal, only by the idea of the type, and not by actual propagation and descent. This is the solution which he sought respecting the "most beautiful organization." And with this he was content.

From Goethe to our contemporary Richard Owen seems a wide leap. But if it was our object to produce in Goethe a stage of natural inquiry which contents itself with a formula of the correlation of living things, dazzling indeed, but ultimately vague, the renowned English comparative anatomist will show us how it is possible to take even the final step and arrive at the conclusion that consanguinity is the sole solution of the similarity of species, and how, nevertheless, by clinging to miracle and dualism, the fruit of the truth just recognized, may be suffered to elude the grasp.[30]

By the personal incitement of Cuvier, under whom he studied in 1830, R. Owen endeavoured to gain a clear perception of the basis of homologies. If Cuvier had derived the agreement of organs from teleology by saying that organs are alike because and if they have like functions to perform, Owen, in Goethe's fashion, seized upon an archetype to explain the existence of uniformity amid multiplicity and diversity of detail. The series which repeat themselves in the organism, such as the vertebræ, and a regular succession in the organisms themselves seemed to him not comprehensible as miraculous creations, but only as the result of natural laws and operating causes, which produce the species in regular sequence and gradual completion, such laws and causes being the servant of predetermining intelligent Will.[31]

As a scholar pre-eminently familiar with the fossil animal world, it could not remain unknown to this English naturalist that the more remote the geological period, the more general and the less specialized is the organization of the species. He was able to trace this particularly in the dentition of mammals, and specially also in the condition of those domestic animals which begin with the earliest Tertiary times and gradually assume the ungulate character. Thus to the question whether species originate by miracle or by law, he replies that he presumes the latter to be in constant operation. This "law" is, however, something quite different from what science is wont to designate by that name. Why does the horse exist? Because it was predestined and prepared for man by the Deity.[32] This is supposed to occur by means of the "derivative law." But this again is a word which conveys no meaning, a phrase which implies that the horse has become a horse because it was so to be. The predecessors of the horse modify themselves for the interest of man, who does not as yet exist, but is already taken into account by the intelligent Will.

These ancestors of the horse might therefore be compared to the sports of Nature; the transformation takes place, not because from inherent reasons it must take place, but because it so pleases the intelligent Will. We must beg to decline such "natural laws" as these. Owen says, "I deem an innate tendency to deviate from the parental type, operating through periods of adequate duration, to be the most probable nature or way of operation of the secondary law, whereby species have been derived one from the other.[33] From the Ichthyosaurus to Man, he sees the connection of descent; he denies that the influence of circumstances is decisive; he rejects a dozen times any sort of miracle; but the next moment he cleaves to miracle again, namely, to an innate tendency towards a certain future development not imposed by circumstances and dependent on them, but conducive to a special purpose.

Thus deal the trimmers, who, through fear of consequences, appease their scientific consciences with a word.

We now come to a courageous writer, whose principal work, "La Philosophie Zoologique,"[34] was overlooked and well-nigh forgotten for half a century, until it was restored to merited honour by Darwin, but more especially by Haeckel, and quite recently in France by Ch. Martins. This is J. B. Lamarck, who first formulated the doctrine of Descent, and in 1804 actually propounded all the propositions which Darwin has constructed afresh and more completely. Lamarck proclaimed that it is merely our limited powers of comprehension that demand the erection of systems, whereas all systematic definitions and gradations are of artificial nature. We may be assured that nature has produced neither orders, families, genera, nor immutable species, but merely individuals which succeed one another, and resemble those from whom they descend. But these individuals belong to infinitely divergent races, which continue so long as they are unaffected by any cause producing alteration. Starting from species, like ourselves, he demonstrates their instability. From comparisons of the facts of hybridization and the formation of varieties, he inferred "that all organizations are true productions of Nature, gradually evolved in the course of a long succession of ages; that in her progress, Nature began, and even now always begins again, with the formation of the simplest organic bodies, and that she directly forms these only, namely, those lowest living beings which have been designated as spontaneous generations."

Variations and transformations supervene, according to Lamarck, through external influences; in the lapse of ages they become essential differences; so that, after many successive generations, individuals which originally belonged to another species ultimately find themselves converted into a new one. The limited period of our existence has accustomed us to a standard of time so short as to give rise to the vulgar and false hypothesis of stability and immutability. The transformation is effected by the obligation of the individual to accommodate itself to the altered conditions of life. Fresh circumstances elicit fresh requirements and fresh activities. Great weight must be laid on the use or disuse of organs. "In every animal still in the course of development, the more frequent and sustained use of an organ gradually fortifies, developes and enlarges it, and endows it with strength proportional to the duration of this use; while the persistent disuse of an organ imperceptibly weakens and deteriorates it, diminishes its efficiency in an increasing ratio, and ultimately destroys it." "And thus," he says, "nature exhibits living beings merely as individuals succeeding one another in generations; species have only a relative stability, and are only transiently immutable."

Lamarck touches upon the struggle of each against all (I. 99, and elsewhere), but does not discover the term Natural Selection. He is fully conscious of the two factors, heredity and adaptation, but his theories and convictions lack the emphasis of detailed evidence. Yet his subtle apprehension of life may be evinced by his interpretation of instinct. According to him, all acts of instinct are effected by incitement, exercised upon the nervous system by acquired inclinations (*penchans acquis*); and these acts, not being the product of deliberation, choice, or judgment, certainly and unerringly satisfy the requirements experienced and the inclinations resulting from habit. But if these inclinations to maintain the habit and renew the actions related to them, are once acquired, they are henceforward transmitted to the individuals by means of reproduction, which maintains the structure and the disposition of the parts in the condition attained, so that the same inclination pre-exists in the young individuals before they put it in practice. This explanation, as Darwin has shown, certainly does not suffice for all the facts of instinct, yet it stands far above the modern "Philosophy of the Unconscious" (Philosophie des Unbewussten), which places the organisms by which the instincts are effectuated, under the sway of an extraneous metaphysical Being who governs it in subservience to design.[35]

VII.

Lyell and Modern Geology--Darwin's Theory of Selection—Beginning of Life.

EVER since mankind has consciously laboured in the field of intellect, pre-eminent men have existed, who, reasoning more rapidly than their contemporaries, have outstripped them in the apprehension of great truths and the recognition of important laws. But it is a great temptation to set too high a value on these anticipations; and in all cases in which these intellectual exploits are concerned, it will be discovered that, so to speak, they floated in the air, and that it was merely a keener scent and a so-called intuition resting on unconscious inferences, which exalted the privileged being above his less sharp-sighted neighbours.

Great scientific crises, revolutions in the domain of intellect, are prepared long beforehand; the watch-word rarely comes too early and is seldom pronounced in accents unintelligible to contemporaries; as a rule, if the change has not been altogether gradual and almost unperceived, but if on the contrary the veil has been suddenly drawn aside by one of these chosen spirits, scales fall, as it were, from the eyes of fellow-labourers and spectators, and the rapidity with which the new theory makes its way affords the best evidence that it took shape and was proclaimed at the proper moment.

That the doctrine of Descent was likewise no utterly startling apparition, even though it leapt forth from the head of Darwin, its greatest representative, like an armed Minerva—of this we have cited at least a few of the many vouchers. That its time had come,—that it was indeed more than time, unless the science of the nature of life, and Biology in general, was to be unduly backward,—is shown by the development of Geology, which thirty years prior to Darwin, after many favourable forecasts, struck upon the right road to the knowledge of causes. The doctrine of the formation and evolution of the earth, especially in its earlier phases, during which Life, in the sense generally attached to the word, originated and became permanent on our Planet,—this science of Geology is intimately allied with our important theme. Modern Geology, especially as connected with the name of Charles Lyell, must sooner or later have necessitated an analogous treatment of vegetal and animal lore, and we can only wonder that the crisis was so long delayed. The exposition of the doctrine of Descent must, therefore, be introduced and initiated by a reference, however brief, to modern Geology.

The first edition of Lyell's "Principles of Geology" appeared in 1830. The tenth, published in 1866, gave him an opportunity of professing his full adhesion to the Darwinian doctrines, to the development of which he had given so great an impulse. Since 1872, the eleventh edition of this masterpiece has been before the world. It treats of the investigation of the lasting effects of causes now in operation, as data from which inferences as to past ages may be drawn. Lyell termed these effects an autobiography of the earth. "The forces now operating upon earth are the same in kind and degree as those which in the remotest times produced geological changes."

Probably, in consequence of the havoc caused by local floods and earthquakes, a belief in great and universal catastrophes was formed at a very early period; and to the Indian and Egyptian legends on this subject Lyell appends the remark, that the traditional connection of such catastrophes with a belief in repeated and universal corruption of morals may be easily explained.

At the end of the last century, the opinion was here and there expressed that the submergence of large extents of land, and the emergence of others, had taken place slowly; and the doctrine was in preparation that the mineral masses fall into various groups, succeeding one another in definite order. Werner then appeared and founded the special science of "Geognosy." He was not the first to see and teach the regular succession of rocks, but the sensation which he caused was universal. From his time dates the violent controversy of the Vulcanists and Neptunists, and into the midst of this controversy fell Cuvier's great discoveries on the animals of the Tertiary formation in the vicinity of Paris. By the works of Cuvier and Lamarck on fossil animals, the differences betwixt ancient and modern organisms became apparent, and Cuvier's views, zoological as well as geological, gained the victory. The conviction was gradually established that long ages of repose and quiescence alternated on earth with shorter periods of universal catastrophes and revolutions.[36]

Even after the appearance of Lyell's "Principles of Geology," the hypothesis of catastrophes received its special completion by Elie de Beaumont's theory of the structure and genesis of mountain chains. From the first, however, Lyell interposed, and derived the following conclusion from a comparison of the slow but continued and perceptible upheavals and subsidences occurring in historic times, with the various modifications which organisms had meanwhile undergone. "In a word, the movement of the inorganic world is obvious and palpable, and might be likened to the minute-hand of a clock, the progress of which can be seen and heard; whereas the fluctuations of the living creation are nearly invisible, and resemble the motion of the hour-hand of a time-piece. It is only by watching it attentively for some time, and comparing its relative position after an interval, that we can prove the reality of its motion."

Careful observation and logical deduction had thus arrived at conclusions diametrically opposite to the assertions of Cuvier, who inferred the geological catastrophes mainly from the striking difference of successive organisms. While botanists and zoologists prosecuted their studies on Cuvier's system, Geology was being metamorphosed under the hands of Lyell and his adherents. He proceeded from the most tangible basis. That it rained during the era of the coal formation, as it now rains, may be seen by the impress of rain-drops on the levels of that formation. The actions of rivers, the sediments of deltas, previously neglected, were now studied, and likewise the colossal mud deposits, such as are exhibited by the Nile and the Amazon, and also the destructive work of the irregular motions of the sea, and the partly destructive, partly formative work of its regular currents. Calculations were made of the ploughing, grating, and grinding of glaciers, of the substances which mineral springs dissolve and deposit, of the displacements of material effected by existing agencies, of the manner in which the outlines of land and sea are altered by elevation and subsidence. Similarly, the comparison of ancient and modern coral reefs and oyster banks showed that these silent builders have not changed their habits. In short, the hypothesis of extraordinary events and forces, unheard of in our present era, seemed quite unnecessary; time only, and the continuous development of the earth's crust, were rendered evident.

The stage for reiterated acts of new creation of organisms had thus collapsed, and the hypothesis of such miraculous new creations became an anachronism, for which a well-merited end was inevitably prepared by the appearance of Darwin. With Darwinism, the doctrine of Descent is an historical necessity.

Charles Darwin was born in 1809, and, as the Naturalist attached to the *Beagle* in her voyage round the world, under Captain Fitzroy, in 1831-7, he enjoyed an opportunity of accumulating rich experiences. His important work on Coral Reefs gave the first adequate explanation of the phenomena resulting from the co-operation of geological movements, and the organic agency of the coral animal; his Monograph on Cirripedes bears witness to the exemplary care with which he can observe and systematically work out the relations of the minutest details. We make this remark, as the opponents of the great inquirer endeavour to suppress his merits and authority by maintaining that he is properly a mere dilettante, dealing with general abstractions,[38] a stranger to the keen observation which takes full account of facts. How Darwin arrived at the idea which has made an epoch in science, he has himself made known in the introduction to his first work on the doctrine of Descent, namely, the "Origin of Species;"[39] and in more detail in a letter to Haeckel, published by the latter in his "History of Creation" (Natürlichen Schöpfungsgeschichte).

"Having reflected much on the foregoing facts, it seemed to me probable that allied species were descended from a common ancestor. But during several years I could not conceive how each form could have been modified so as to become admirably adapted to its place in nature. I began, therefore, to study domesticated animals and cultivated plants, and after a time perceived that man's power of selecting and breeding from certain individuals was the most powerful of all means in the production of new races. Having attended to the habits of animals and their relations to the surrounding conditions, I was able to realize the severe struggle for existence to which all organisms are subjected; and my geological observations had allowed me to appreciate to a certain extent the duration of past geological periods. With my mind thus prepared I fortunately happened to read Malthus's "Essay on Population;" and the idea of natural selection through the struggle for existence at once occurred to me. Of all the subordinate points in the theory, the last which I understood was the cause of the tendency in the descendants from a common progenitor to diverge in character."*

That organisms are variable and not fixed in rigid forms, is a phenomenon so general that variability passes current as a self-evident property of organic existence. In the next chapter we shall inquire how far everything organic is necessarily subject to mutability. On the existence of this property rests the artificial breeding, or selection by man, consciously and unconsciously exercised from the earliest commencement of hunting and agriculture, of which, as Darwin says, "the importance mainly lies in the power of selecting scarcely appreciable differences, which are nevertheless found to be transmissible, and which can be accumulated until the result is made manifest to the eye of every beholder." In the "Origin of Species," as an example of methodic selection in the production of breeds, Darwin has chosen the pigeon, to the breeding of which he zealously devoted himself for many years.

The pigeon is specially adapted to the purpose of scientific observation of the phenomena of breeding, because, owing to its monogamic habits, it is easy to control, because it may be brought in a short time to striking variations, because the records of its breeding are tolerably complete, and, finally, because it is one of the few domestic animals of which the ancestral stock is scarcely open to a doubt.

The chief races produced by the fanciers may be grouped as follows. The Pouter Pigeons have a moderate beak, elongated legs and body, their œso-

* Mr. Darwin has himself been good enough to re-write his letter from the German text. He kept no copy of the original M

phagus is of great size, barely separated from the crop, and is capable of inflation. A second group includes Carriers, Runts, and Barbs, which possess in common a long beak, with the skin over the nostrils swollen and often carunculated or wattled, and the skin round the eyes bare and likewise carunculated. To another group, with shorter beak, and the skin round the eyes only slightly developed, belongs the Fantail, in which the normal number of twelve tail feathers may rise to forty-two with aborted oil-gland; also the Tumbler, in which the beak becomes extremely short, and a sickly disposition of the brain, produced and exaggerated by selection, and manifesting itself by tumbling, has been transmitted for more than 250 years, and has become established as the characteristic of a race. In the fourth group, the Trumpeter occupies a prominent position, on account of its peculiar voice; likewise the Laugher, or Indian turtle-dove, comprising several sub-races scarcely differing in structure from the rock-pigeon (Columba livia). The latter is divided into several geographically distinct races, ranging from the coasts of the Faroe Islands and Scotland to the shores of the Mediterranean and to India; and the most minute investigation, whether the incredibly divergent races of domestic pigeons are derived from eight or nine wild species or solely from the wide-spread rock pigeon, results decidedly in favour of the latter alternative. Proportional dimensions, colouring, and parts of the skeletons which differ from one another far more widely in the various races than they do in well marked species of the same genus, or even family, are modified under the hand and according to the will of man; and, moreover, pre-eminently in the pigeon may be traced the phenomenon which has been termed the "correlation of growth," and consists in the fact that, with the intentional modification of an organ by means of selection, one or more other organs are drawn into sympathy and unintentionally transformed into characteristics of a race.

Darwin's minute researches on the formation of races in the pigeon are recounted in his second work on the theory of Descent, "The Variation of Animals and Plants under Domestication," in which the most detailed investigations respecting other domestic animals are also to be found. Whoever has had occasion to inspect one of the modern exhibitions of poultry, must have been astonished at the diversity of the different races, and the purity and uniformity within each race. Though not quite so positively as in the case of the pigeon, yet with approximate certainty, the domestic fowl appears to be derived from a single ancestral stock, the Indian Gallus Bankiva. The cumulative power of selection by man is likewise testified by the various races of pigs bred within the last century by the English farmers from an intermixture of the native and Indian races, differing in general appearance, colouring, size of ears, length of legs, and also partially in fertility. Our attention is, however, more closely drawn to the two races of Southdown sheep and Shorthorn cattle, which, as well as the choicest breeds of pigs, have been for some years past particularly esteemed on the continent. These and many other races have been bred with definite purposes, and for certain domestic and commercial advantages, and one and all bear testimony to the plasticity of species.

Artificial selection operates by establishing peculiarities

originally variable, and on their first accidental appearance usually perceived only by the careful eye of a connoisseur. But not a few cases are likewise certified in which an accidental deformity and a new character appearing suddenly even in a single individual have lent themselves to the rapid formation of a race. "Thus," as Darwin relates,[41] "in 1791 a ram lamb was born in Massachusetts, having short, crooked legs and a long back like a turnspit dog. From this one lamb the otter or ancon semi-monstrous breed was raised; as these sheep could not leap over the fences, it was thought that they would be valuable; but they have been supplanted by merinos, and thus exterminated. These sheep are remarkable from transmitting their character so truly, that Colonel Humphreys never heard of but one questionable case of an ancon ram and ewe not producing ancon offspring."—"A more interesting case has been recorded in the Report of the Juries for the Great Exhibition (1851), namely, the production of a merino ram lamb on the Mauchamp farm in 1828, which was remarkable for its long, smooth, straight, and silky wool. By the year 1833, Mr. Graux had raised rams enough to serve his whole flock, and after a few years more he was able to sell stock of his new breed. So peculiar and valuable is the wool, that it sells at 25 per cent. above the best merino wool; even the fleeces of half-bred animals are valuable, and are known in France as the Mauchamp merino. It is interesting, as showing how generally any marked deviation of structure is accompanied by other deviations, that the first ram and his offspring were of small size with large heads, long necks, narrow chests, and long flanks; but these blemishes were removed by judicious crosses and selection. The long, smooth wool was also correlated with smooth horns; and as horns and hair are homologous structures, we can understand the meaning of this correlation. If the Mauchamp and the ancon breeds had originated a century or two ago, we should have had no record of their birth, and many a naturalist would, no doubt, have insisted, especially in the case of the Mauchamp race, that they had each descended from or been crossed with some unknown aboriginal form."

If with the refined culture of races on large estates, we compare the slight attention bestowed on domestic animals in small peasant farms, remote from the cheering intercourse of the world, and then descend to the treatment by savages of their few domestic animals, or their sole tame creature, the dog, conscious artificial selection gradually decreases; but wherever man attaches to his abode either plants or animals, selection is at least unconsciously exercised. The powerful animal, the particular plant which yields the most abundant nutriment, are employed for propagation without any special forethought, and unconscious selection is thus undistinguishable from that which is methodically practised. The initiation and progress of the production of races is naturally facilitated by the power of placing the animals selected for breeding, in a new environment and fresh conditions of life, and the formation of new races is favoured by the ease with which it is possible to hinder the crossing of forms in course of construction, with races already existing.

Unquestionably many races of domestic animals are not in a condition in which they can be termed new species; that is to say, in regard to the new characters evolved by breeding, they are in a state of merely artificial stability; and, if abandoned to accidental or irregular intermixture with the aboriginal or other races, they gradually revert to their primitive form. But it is arbitrary and erroneous to assert that all unconsciously or consciously bred races, without exception, are no new species, and would all relapse if left to a state of nature.

Granting that all the races of fowls were left to themselves, we must certainly admit the possibility that in India some few forms would change back into the Bankiva fowl. It is, however, evident that, in Europe and America, from any semi-feral races of fowls the aboriginal Indian race would never reappear, but at the most some few new wide-spread mongrel forms would arise, remaining constant according to geographical districts. No one has yet been able to assert that the wild dogs of the East, entirely released from the control of man, have become wolves or jackals, their presumptive ancestors. They become "jackal-like," by which every one expresses that the dog which became and was bred a domestic animal thousands of years ago, preserves its acquired specific characteristics even under circumstances most favourable to their destruction.

This statement, that domestic animals are no new species, is the more unfounded, as of several domestic animals the aboriginal stock is totally unknown; among these are the sheep and goat, respecting the ancestors of which only vague conjectures can be framed. The most ancient race of sheep known to us,—that with ram-like horns, found among the lake dwellings of Switzerland, throws no light upon the subject; and empirically to observe the reversion of the modern sheep to its aboriginal form is utterly impossible. That the horse is derived from a striped aboriginal species is probable; but notwithstanding the many generations during which the great herds of feral horses in South America have propagated themselves undisturbed, no such species has been produced. Rütimeyer's minute researches on domestic cattle have shown that, in Europe at least, three well-defined species of the Diluvial period have contributed to their formation, Bos primigenius, longifrons, and frontosus. These species once lived geographically separate, but contemporaneously; and they and their specific peculiarities have perished, to rise again in our domestic races. These races breed together with unqualified fertility; in the form of skull and horns they recall one or other of the extinct species; but collectively they constitute a new main species. That from their various breeds, the three or any one of the aboriginal species would ever emerge in a state of pristine purity, would be an utterly ludicrous assertion.

In all these domestic animals—dog, sheep, goat, horse, and cattle—the transformation was initiated in an era of civilization in which there was no idea of artificial breeding in the modern sense, and in which the main factor of transformation, independently of involuntary and unconscious selection, consisted simply in the altered mode of life. This introduces us to variations in a state of nature, and to Natural Selection. Natural as well as artificial selection both rest on the undisputed fact of the idiosyncrasies of the most closely allied vegetal and animal individuals; and it has already become manifest that doubtful species are not exceptional, as the old school was wont to imagine, but that it is merely owing to an inadequate knowledge of the material out of which species are constructed, that all species are not looked upon as doubtful and artificial.

Let us here again call to mind that in many thousand cases the most rigid systematizers are unable to state where their species begin and end; of which Darwin, as an instance, cites a communication by H. C. Watson, that 182 British plants, usually regarded as varieties, have each been claimed as independent species by individual botanists.[42] Darwin's immortal service consists in having shown what is the power which operates upon the existing variable individuals and species, and what results this operation must produce. He found the key in the word which has become a badge and common property of our age, "the struggle for life,"* and has thus given the foundation and theory of a doctrine of which the truth had long before been manifest to an intellect such as that of Lamarck. He founded the doctrine of Descent on the theory of selection, when he proved that in nature the struggle for existence occasions a selection of the best and fittest, comparable to artificial breeding, and giving rise to new races and new species.

The struggle for life, this *bellum omnium contra omnes*, is, moreover, an undisputed and undeniable fact, which we here accept in its widest relations. Not only does the beast of prey war against the graminivorous animals, which again strive to keep their balance by superior multiplication, speed, and cunning; the gradual

* For Wallace's share in this honour, see the end of this chapter.

advance of a plant is likewise a struggle with natural obstacles; and the conquest which it gains usually injures other plants in their conditions of life. If the powers of multiplication of any given organism were to operate absolutely and unrestrictedly, each being would, in a short series of years, claim for itself the whole surface of the earth, or all the waters of the sea. But each holds the other in check; and with the living foes of each creature are associated the climate and all the influences of the surrounding conditions, and of the alternation of the seasons, to which the body must accommodate itself. Organisms live only at the cost of, and for the profit of, others; and the peace and quiet of nature sung by the poet is resolved under the searching eye into an eternal disquiet and haste to assert and maintain existence, amid which it is only the thought of the visible and necessary progress that can rescue the observer from a pessimist view of the world.

The simplest examples of the relations of mutual dependence of living beings are, however, the best and most conclusive; but the vast consequences depending on circumstances and connections apparently insignificant, and the extreme complexity of the mechanism by which equilibrium is maintained, have been exhibited by Darwin in some examples, which, frequently as they have been repeated, we shall also allow ourselves to reproduce. Whereas, to the South and North of Paraguay feral cattle, horses, and dogs abound in profusion, they are wanting in Paraguay itself. "Azara and Rengger have shown that this is caused by the greater number in Paraguay of a certain fly, which lays its eggs in the navels of these animals when first born. The increase of these flies, numerous as they are, must be habitually checked by some means, probably by other parasitic insects. Hence if certain insectivorous birds were to decrease in Paraguay, the parasitic insects would probably increase, and this would lessen the number of the navel-frequenting flies; then cattle and horses would become feral, and this would certainly greatly alter (as indeed I have observed in parts of South America) the vegetation, and this, again, would largely affect the insects, and this the insectivorous birds, and so on, in ever-increasing circles of complexity."

Another example out of Darwin's store is perhaps even more striking. "I find from experiments that humble-bees are almost indispensable to the fertilization of the heartsease (Viola tricolor), for other bees do not visit this flower. I have also found that the visits of bees are necessary for the fertilization of some kinds of clover; for instance, 20 heads of Dutch clover (Trifolium repens) yielded 2,290 seeds, but 20 other heads, protected from bees, produced not one. Again, 100 heads of red clover (T. pratense) produced 2,700 seeds, but the same number of protected heads produced not a single seed. Humble-bees alone visit red clover, as other bees cannot reach the nectar. It has been suggested that moths may fertilize the clovers; but I doubt whether they could do so in the case of the red clover, from their weight not being sufficient to depress the wing-petals. Hence we may infer as highly probable that, if the whole genus of humble-bees became extinct or very rare in England, the heartsease and red clover would become very rare or wholly disappear. The number of humble-bees in any district depends in a great degree on the number of field-mice, which destroy their combs and nests; and Colonel Newman, who has long attended to the habits of humble-bees, believes that more than two-thirds of them are thus destroyed all over England. Now, the number of mice is largely dependent, as every one knows, on the number of cats; and Col. Newman says, 'Near villages and small towns I have found the nests of humble-bees more numerous than elsewhere, which I attribute to the number of cats that destroy the mice. Hence it is quite credible that the presence of a feline animal in large numbers in a district might determine, through the intervention first of mice and then of bees, the frequency of certain flowers in that district.'"

The closer the kindred of the competitors, the more ardent is the struggle for the existence; for the more adjacent organisms differ in their requirements, the less do they interfere with one another, and the more will each be able to exhaust the resources of the vicinity for its own benefit. This seems to be flatly contradicted by the great series of associated plants and animals; but on closer inspection they also form no exception to the rule, as, often by their very number they render existence possible and easy to one another, and increase exactly in the degree permitted by the stock of nutriment. If among associated plants or gregarious animals a surplus production occurs, competition and conflict instantly commence, and life is regulated in every respect exactly as in species less remarkable for the number of individuals.

Our proposition that the vehemence of the struggle rises with the closeness of the kindred, is thus universally valid. Such a rapid war of extermination is rarely waged as that between the black rat (Mus rattus) and the brown rat (Mus decumanus); and we far more frequently imagine that harmonious intercourse exists between the members of the same species sharing the same habitation, as, for instance, the hare and the deer, than that they are anxiously striving to maintain existence. Yet this is not the case. The two great motive powers, the preservation of the individual and the preservation of the species, are unremitting instigations to warfare, and under their influence every living being, plants inclusive, joins in conflict with its congeners of the immediate vicinity.

In this competition for nutriment, combined with defence against all possible enemies and other rivals for the remaining privileges of existence, the strongest gains the advantage, or the most crafty, the most skilful—in short, the one that can measure itself against its rivals armed with any sort of superiority. Not only in the struggle for mates, but on every occasion of competition, the weaker individuals are beaten off, and a selection of the strongest and the best takes place. But the primarily slight,—often scarcely perceptible, advantages, mental as well as bodily, which aided these individuals to conquer and survive the other members of the species who were weaker and destitute of accidental advantages, have a prospect of being transmitted, and in the following generations of becoming established and increased by repeated selection. This selection is therefore a natural and necessary course of things; and it applies, not in a merely general and vague manner, as in the external habit, size, and strength of the individual, but, owing to the actual variability and plasticity of the organic morphological constituents, single parts and organs may be modified and perfected in definite advantageous directions, so as to secure for the race and species a higher position in the surrounding world.

Besides the general results of the right of the strongest, another very influential phenomenon comes into play where the desire for propagation is concerned, which Darwin has designated as "sexual selection," and elaborated in great detail in his work on the "Descent of Man." In this we must consider, first, the formation of sexual peculiarities in the males, and the secondary characters by which they are aided in the courtship of the females; and only secondly the reactions of these peculiarities on the alteration and progress of the species in general.

The fundamental idea of Darwin's theory of selection is therefore, that the cumulative power of selection exercised by man in the breeding of races, is, in nature, replaced by the struggle for life; and that in the course of time, by the cumulation of advantages primarily slight and becoming more and more prominent, lower organisms are converted into higher ones. The process is incessant. "It may be metaphorically said, that natural selection is daily and hourly scrutinizing throughout the world the slightest variations, rejecting those that are bad, preserving and adding up all that are good; silently and insensibly working, whenever and wherever opportunity offers, at the improvement of each organic being in relation to its organic and inorganic conditions of life." [44]

The following chapters will introduce us more nearly to this theory, its truth, possibility, application, and confirmation; meanwhile we will at once make ourselves acquainted with some of the objections offered to it, either to the theory of selection in particular, or to the theory of selection combined with the doctrine of transformation as a whole; the most important of which Darwin has already considered and answered.

If, so it is said, all living beings stand in distinct and uninterrupted connection with one another, what has become of the infinitely numerous intermediate forms which must necessarily have existed? Our eyes turn first to the organisms now living, and as, in accordance with the theory, they are assumed to be the terminal twigs of an infinitely ramified tree, which must obviously press hard upon one another, and must each independently diverge in all directions as varieties, we ask for the intermediate forms of the species now existing side by side.

We may now appeal to the evidence already given (p. 92, &c.), that in complete and extensive groups of organisms, modern scientific research has been able to discern nothing else than intermediate forms. Similarly, the journey undertaken by Kerner in his little book on "Good and Bad Species," in company with the botanist Simplicius, from the West of Europe to the East, will furnish an amusing number to the reader eager for further material. The extension of the various species of Cytisus which this naturalist has minutely investigated, likewise exhibits the uninterrupted existence of connecting forms on the territorial boundaries of species of which the centres of propagation are more or less remote. From all these instances, which may be reckoned by thousands, it may be inferred that a large proportion are in a phase of relative stability. That for this reason their intermediate forms must be looked only for in the past, is as little surprising; it in no way impugns the truth of the doctrine of Descent; and the demand for intermediate forms between these local and temporarily stable forms merely proves how little those who make it have appreciated the nature of Descent.

But the objection mainly concerns those intermediate forms by which the species are connected with the aboriginal species preceding them in order of time. According to the theory, the species now living are connected with the aboriginal species by forms identical in quality with varieties, the "species in process of formation;" the aboriginal species with others still more ancient, and so on; so that an infinite number of forms must have existed. We have already shown (p. 97, &c.) that in an excess of zeal palæontologists have set up species, also to be reckoned by thousands, where merely transitional forms and varieties actually existed; we have mentioned that a number of distinguished palæontologists of the present day are endeavouring to remedy the errors of their predecessors, and to exhibit the uninterrupted transitional series from the lower to the more recent strata, where the others with lavish ingenuity imagined they had discerned specific characters. Still it must be admitted that the amount of transitional forms as yet actually found are a vanishing quantity, as compared with the countless multitude which must have existed.

But this deficiency may be satisfactorily explained. We know only a very small proportion of the fossiliferous strata, and, with as much justice as Lamarck in the beginning of this century, we may point even now to the poverty of the collections. Wherever the palæontologist now lays his hand he finds intermediate forms, and day by day the material accumulates as it is required. Nevertheless, too much is demanded, and the conditions of preservation are misunderstood, if it be supposed that all intermediate forms that ever existed were, by their bodily constitution, either wholly or partially adapted to preservation, and must therefore have been actually preserved. On the contrary, the greater number have assuredly vanished without a trace. At least half of all geological deposits have been destroyed again during slow upheavals. For from the time at which a sea-bottom formerly lying at a profound depth, with its well-preserved enclosures, is again raised within the reach of superficial movements, it may be crumbled and corroded, and the fossils contained in it now share the fate which usually befalls the remains of the denizens of marshy shores, they are triturated by the surf.

To this must be added the important consideration that the forms by which the transition is effected will mostly, not as individuals, but as forms, have had a briefer period of existence than the persistent varieties appearing to us as species, as may be seen, among other instances, in the instructive discoveries at Steinheim. In this particular, the periods of transition from one geological plane to the next, resemble the boundary regions of two geographical districts. The tract of transition from one to the other is specially suited to give rise to the transformation of appropriate organisms. But this transformation is accomplished and established first in the new district. Thus in the geological series, transitional periods are periods of relative disturbance. During their continuance the exigency of adaptation and transformation was at its height, both in the vegetal and animal world; the conditions of existence at the same time most unfavourable; the number of individuals in those species which succeed in effecting their transformation is necessarily reduced, and could increase again only in the subsequent periods of repose. It is therefore not surprising that the catalogue of intermediate forms is so defective; and their scarcity is remarked upon only by those who are determined to feel the want of them. For the establishment of scientific evidence in favour of the doctrine of Descent we have them in superabundance.

With the supposed deficiency of transitional forms is connected another frequent objection, namely, that in repeated instances whole groups of kindred species have suddenly appeared. If intermediate morphological and anatomical gradations are elsewhere visible, in these groups, the pterodactyls, birds and others, there is no coherence and connection with any aboriginal species previously or contemporaneously existing. This allegation is one of the feeblest and most vapid, if raised after the attempt has been made to account for the absence of intermediate forms. It is only a particular case in the alternative that either all species originated in the natural manner indicated by the perfectly adequate number of transitional forms at hand, or all by miracle. In the cases which are here brought forward as heavy artillery, the gap to the aboriginal species is certainly greater than where there is merely a leap from species to species or to genus. But the explanation given of the less striking intervals scarcely needs extension to suffice here also. The obscurity overshadowing the extraction of the birds is just beginning to clear up. Why should not the origin of pterodactyls become more distinct in the next few years?

A special difficulty is seemingly prepared for the theory by the highly integrated organs, particularly by the apparatus of the senses, with their very complex mechanism. In truth, taking, for example, the eye of the Vertebrata, we must not even say of the higher Vertebrata alone, that its marvellous structure is well fitted to excite the liveliest doubts as to descent and selection. As a matter of fact, however, the series of vertebrate animals does not exhibit the series of lowly beginnings which we must assume as having once existed. For the eye of the fish is little inferior in complexity to the optic organ of the mammal, whilst the lancelet is completely eyeless, and therefore affords no clue.

In other orders of animals, however, we still see in the systematic series of the present era every possible gradation, and thus possess a representation of the manner in which in the palæontological series the perfect organ was gradually evolved from the simplest rudiments. The lowest crabs present the simplest mechanism imaginable, sensitive to light; other crabs of higher development possess eyes somewhat more perfect, not only sensitive to light, but capable of forming images, and between these eyes and those of the decapodous crab, so extremely perfect of their kind, a host of optic structures are represented, which clearly show that these organs are also subject to the law of slow accumulation and establishment of small advantages.

With regard to the auditory and olfactory apparatus,

every manual of comparative anatomy affords testimony that the series of vertebrate animals now living, present series of development which negative the sudden and incomprehensible origination of these organs in an immediate state of completion. How they appeared in yet lower grades than are exhibited in the true fishes of the present time we may learn in part from the lancelet, and in part we may picture to ourselves from the corresponding sensory apparatus of the lower Mollusca, Articulata, and Annulosa. With reference to the objections to his doctrine arising from the arrangements of the most perfect organs, Darwin has said that he would abandon his whole theory if it can be shown that any of these organs could not possibly have been formed from lower grades, by improvement slowly acquired. This demonstration no one has yet undertaken, nor will it ever be undertaken with success, as every deeper penetration into the comparative anatomy of the sensory apparatus affords evidence to the contrary. In order to understand the presumptively faultless sensory organs and their derivation from a lower grade, it is of supreme importance to bear in mind the circumstance first exhibited by Helmholtz in the eye, that besides a number of perfections, they likewise possess a number of imperfections, and purposeless or obstructive arrangements.

But we must examine another point, which may awaken doubts as to the admissibility of the doctrine of Descent, though, strangely enough, it has as yet been turned to little account by adversaries, and only incidentally touched upon by Darwin. In the "Origin of Species," he states, that H. C. Watson, we know not where, has opposed to the Divergence of Character, or the inclination of varieties and species to deviate from one another, a Convergence of Character. It is conceivable, he thinks, that species derived from different genera, might sometimes approach each other so closely that they would be classed in the same genus. The author of the theory of Selection has been content to point out the great improbability of such an event, which, moreover, in this simplicity would scarcely impugn the origin and truth of the theory. "If two species of two allied genera both produced a number of new and divergent species, I can believe that they might sometimes approach each other so closely that they would, for convenience' sake, be classed in the same new genus, and thus two genera would converge into one; but from the strength of the principle of inheritance, and from the two parent species already differing, and consequently tending to vary in a somewhat different manner, it seems hardly credible that the two new groups would not at least form different sections in the same genus."[45]

We here see a theoretical objection theoretically refuted. But although the probability of a convergence carried to absolute similarity is extremely slight, and it receives no support from the palæontological record, its utter *à priori* impossibility must not be rashly asserted; and in my researches on the Sponges of the Atlantic, I have pointed out groups of species approximating so closely as to be scarcely distinguishable. Chalina and Reniera are two distinct genera, actually belonging to different families. It is highly probable that the genus Chalinula, with its extremely variable species, are branches of the Chalina, not the converse; and the forms of Reniera likewise merge in species not constant in any character, and which the most careful observer is unable to distinguish from the Chalinula. Therefore, if the convergence or approximation of branches of various origin cannot be rejected in the abstract, the most propitious case of coincidence is, nevertheless, limited to the province of analogous formations, where, under like conditions of adaptation, different families have been driven to like expedients and differentiations, producing complete similarity. A general survey of the organic world likewise teaches us that in the higher regions this overlapping of the ends of dissimilar parentage becomes more and more incredible, and, as is shown by my study on sponges, they can, in any case, occur only where the organisms consist of very simple factors, highly variable in a few directions, and very easily affected by external conditions.

When we referred above to the possibility of no slight objection to the doctrine of Descent, we spoke of another case of convergence. We mean, namely, those similar final results in divergent series by which, in highly organized groups of animals, of which the reciprocal connection can be traced only through low aboriginal forms, certain important organs exhibit the greatest uniformity of arrangement and integration. It is, as yet, quite undecided where and when the true insects separated themselves from the water-breathing crabs; nay, some naturalists incline to the opinion that these two classes are derived from a more remote common ancestor. Thus much is extremely probable, that the severance between crabs and insects took place when the development of their optic apparatus had not attained the degree of perfection which we now find in the stalk-eyed crabs and insects. They nevertheless agree not merely in their coarser conditions, but, as Max Schultze has demonstrated, even in their minutest microscopic details. If the idea of design as a principle of explanation is excluded in this case also, as will be shown below, and as is self-evident from our standpoint, and if simple heredity in the two series must be excluded likewise, some other adequate solution must be sought.

The case of the converging species of sponges may throw a light, feeble though it be, upon the obscure processes of the organic laboratory. Let us here again recall that maxim of Goethe, which we have already cited: "The animal is formed by circumstances for circumstances." Perhaps this maxim may in future be brought into play, for it is actually a question of investigating how surrounding conditions, the agencies acting on the sensory apparatus, can exercise on simple matter such an influence, that the otherwise widely differing descendants of the various possessors of this simple material or incomplete organs, have acquired a more complete organ, not only working in a similar manner, but of similar construction. Darwinism has never yet pretended to have explained everything; neither will it be wrecked on this point, but, on the contrary, will only have supplied fresh incitements to more profound researches, crowned by beautiful results.

Another example of approximation in divergent series is afforded by the eyes of the highest molluscs, the Cephalopods, as compared with those of the Vertebrata; in this instance, however, it does not go beyond an analogy, though a striking one. It is only the microscopic structure of the nerve membrane, which is extremely similar in both divisions, with the exception of the reversed sequence of the layers from inwards, outwards. The case, considered in the abstract, appears highly complicated, and without a prospect of solution; but it becomes marvellously simplified, as we have already hinted, if the question is thus generalized: In what manner are the still undifferentiated terminations of the nerves affected by the specific operation of the waves of light and sound, &c., so as to assume the form and construction of specific peripheral organs? It may be long before these relations are fathomed; our only concern is to defend the theory from the reproach of inadequacy, by showing the scope for investigation according to our point of view.

When Darwin had brought to light the effects of natural selection in reproduction and derivation, and applied this principle to all the phenomena of the organic world, the systematic school was effectually subdued by that same doctrine of Descent, thus fortified and established, after which Lamarck had striven in vain. The systematic school classified organisms according to external and internal resemblances. Whence this greater or smaller accordance, whence the gradation and the heterogeneity, it knew not how to tell. It was thought that much was gained when the fundamental forms of types were spoken of, even though no account was given of the intrinsic nature of these types, floating like ideas above the phenomena. Now the type has become the family, and the systematizers have the plain task of restoring and combining the pedigrees of the various groups of living beings. The knowledge of these pedigrees has now for the first time a truly scientific purport, as compared with the old system of types; for the genealogical trees cannot be constructed without a knowledge of their growth, and of the causes which produced their branches, twigs, and shoots. Each family thus includes all the forms derived from one simple original form. The old systematic school was obliged to content itself with working out the classification of the individual types, and defining their limits, and then balancing the types against each other on general morphological and physiological principles, in order to estimate their relative value, all without any consciousness of the natural causes of these actual relations. The doctrine of Descent connects the original forms of the types afresh from the point of view of consanguinity, and descends deeper and deeper, down to the simplest organisms, and the beginning of life.

But before we attempt to come to an understanding as to the origin of life, one of the pillars of the doctrine of Descent, it seems appropriate to allude to the question whether natural selection, of which the means and effects will be more minutely elucidated in the following chapters, is capable of explaining all the modifications of organic beings, and whether selection must always be summoned to aid in the explanation of these transformations? In other words, whether the theory of selection answers all the requirements of the doctrine of Descent, or whether it is capable and in need of amendment? We may do this with the more impartiality, as the acute author of the book entitled "The Unconscious from the Standpoint of Physiology and the Theory of Descent (Das Unbewusste vom Standpunkt der Physiologie and Descendenztheorie),[46] has again recently observed that the truth of the doctrine of Descent is independent of the bearings and adequacy of the Darwinian theory.

"This circumstance," he says, "is misunderstood by the majority of Darwin's opponents; when they adduce arguments for the inadequacy of natural selection in the struggle for life, they usually fancy they have adduced just as many arguments against the reliability of the theory of Descent. But the two have no direct connection with one another; for it might be possible that Darwin's theory of natural selection was absolutely false and unserviceable, and the doctrine of derivation true notwithstanding; that only the causal medium of the derivation of one species from another was different from that stated by Darwin. Similarly, it might be possible that, although the mediate causes of transition discovered by Darwin were partially effective,—on the other hand, transitional phenomena existed which could not as yet be explained by this hypothesis; that this therefore required either an auxiliary hypothesis supplementing that of Darwin, or even a co-ordinating principle of explanation, as little discovered now as was the Darwinian theory twenty years ago. Such imperfect knowledge of the causes operating in the transition of one form into the other, can prejudice the general truth of the doctrine of Descent as little as the absence of intermediate forms, or the uncertainty, still existing in many cases, of the derivation of any given form. If even in former times, when all knowledge of the causes by which transition is effected was still wanting, the doctrine of Derivation seemed certain to the greatest minds, on philosophical and *à priori* grounds, there can be still less doubt as to the theory of Descent, now that Darwin and Wallace have plainly shown that indubitably the most important, if not the all-sufficing cause of transition is everywhere effective, and in many cases sufficient."

We wished to set forth these words of a talented philosopher for the benefit of those who are so unreasoning as to pour away the child with the bath-water, and fancy that they have slain the doctrine of Descent when they have been lucky enough to raise a few cavils against Darwin's theory of selection. Does the theory of selection fulfil every requirement? It accomplishes many and great things, but in some cases it seems to be inadequate, and in other cases it is not requisite, as the solution of the formation of species is found in other natural conditions.

Moritz Wagner, a decided adherent of Metamorphosis and an enthusiastic admirer of Darwin, endeavoured to establish a "law of migration," namely, that "the migration of organisms and the formation of colonies by them is the necessary condition of natural selection."[47] In his opinion, new species arise only when smaller communities of individuals, in process of forming varieties, are geographically isolated, as in this manner only is intercrossing precluded with their stationary congeners, who do not participate in the transformation; and reversion and disappearance of characters as yet not fixed is thus avoided. That isolation often acts very favourably on the formation of species is a fact almost universally acknowledged and easily verified by insular fauna, but that the formation of species can take place only with the assistance of isolation has been effectively refuted by Weismann.[48] He has shown that an "inter-crossing of the incipient variety with the aboriginal form is not avoided by isolation;" and by the very favourable instance of the lake of Steinheim, among others, he has exhibited the formation of new species in the midst of the old ones. On Haeckel's remark that in the asexual propagation of the lower beings, the influence of inter-crossing was not to be feared, Wagner had already restricted the necessity of isolation to the higher organisms with separate sexes. But Weismann most justly insists that Wagner's "law of migration" is deprived of all foundation by one of the most remarkable examples of the formation of varieties on the same territory, namely, the fact of the separation of the sexes, as to the derivation of which from species once hermaphrodite, all (the believers in Creation naturally excepted) are assuredly of one accord.

As we have already mentioned, it seems that if the impulse to form varieties once exists, the tendency spreads rapidly. Steinheim, with its Planorbis multiformis, is specially propitious to the demonstration of these periods of variation. If isolation coincides with such a period, it effects the establishment of new varieties into species without the aid of natural selection. As Darwin admits in his work on the origin of Man, he formerly bestowed too little attention on the formation of so-called morphological species. By this we mean, species not distinguished from their aboriginal stocks by any physiological advantages, and hence not superior to them, in which therefore the principle of selection in the strict Darwinian sense is inapplicable. Two species of butterflies, differing only in a few specks or pencilings, or the notches on the wings, are in our estimation of perfectly equal physiological value; they are morphological species. Weismann sets up the proposition that "the colouring and penciling of the upper surface of the wing in butterflies are to be regarded as purely morphological characters, excepting in cases of mimicry and protective uniform colouring." He shows also by other examples that, "under certain circumstances and within a comparatively small range, new as well as morphological characters may be established by the effects of isolation only." The inapplicability of natural selection to the evolution of purely morphological variations was first pointed out by Nageli.[49] With reference to this subject, Darwin with magnanimous modesty observes: "I now admit, after reading the essay by Nageli on plants, and the remarks by various authors with respect to animals, more especially those recently made by Professor Broca,[50] that, in the earlier editions of my 'Origin of Species,' I probably attributed too much to the action of natural selection or the survival of the fittest. I have altered the fifth edition of the 'Origin' so as to confine my remarks to adaptive changes of structure. I had not formerly sufficiently considered the existence of many structures, which appear to be, as far as we can judge, neither beneficial nor injurious, and this I believe to be one of the greatest oversights as yet detected in my work."[51]

We are disposed to think that the oversight with which Darwin charges himself is not so great, as it is here a question of the more indifferent species, not affecting the great phenomena of progressive development, and of which the origin is perfectly comprehensible by variability alone, or in any case, as we have seen, by variability with the co-operation of isolation. The value of natural selection is in no way deteriorated by the possibility of explaining the purely morphological species without its aid. In certain cases of mimicry, or the formation of natural protective masks and imitations, and for the explanation of organic beauty, natural selection seems inadequate. But what does this prove, but that, as all know, future generations must needs carry on the edifice? The additions which the presence of the theory of selection has been able to supply are scarcely worthy of mention.

As the type has become the family, and the system, as the shortest expression of the kindred relations of organisms, requires at the root of the genealogical tree a number of the lowest and simplest organisms, or perhaps one single primordial form, we must come to an understanding as to the problem of the beginning of life. Even quite recently, in March 1873, Max Müller, in accordance with an opinion shared by many, has again proclaimed "the Darwinian theory vulnerable at the beginning and at the end."[52] Whether any considerable points of attack are offered by the final proposition of Darwinism, namely, the application of natural selection to man, and his sole characteristic peculiarity, language, we shall have another opportunity of inquiring. But what the renowned linguist terms the vulnerable beginning of Darwinism, the origin of life, has in fact nothing to do with actual Darwinism, or natural selection, unless the principle of selection be extended to the inorganic world of matter. But the objection which endeavours to cut away the ground from under the doctrine of Descent, not the theory of selection, and represents the origin of life as incomprehensible and supernatural, we naturally regard as an attempt to gain a precedent for the supernatural creation of language. Between beginning and end, we naturalists may do as we please.

But it is strange that the very side which is so ready to reproach us with a want of philosophic method and induction, should here, where the material substratum is deficient, dispute the claims of the investigation of nature to its logical inferences. In the last page of the "Origin of Species," Darwin says: "There is grandeur in this view of life, with its several powers, having been originally breathed by the Creator into a few forms or into one; and that, whilst this planet has gone cycling on according to the fixed law of gravity, from so simple a beginning endless forms most beautiful and most wonderful have been and are being evolved." In this concession, Darwin has certainly been untrue to himself; and it satisfies neither those who believe in the continuous work of creation by a personal God, nor the partizans of natural evolution. It is directly incompatible with the doctrine of Descent, or, as Zöllner[52] says: "The hypothesis of an act of creation (for the beginning of life) would not be a logical but a merely arbitrary limitation of the causal series against which our intellect rebels by reason of its inherent craving for causality. Whoever does not share this craving is beyond help, and he cannot be convinced. To hold the beginning of life as an arbitrary act of creation, is to break with the whole theory of cognition."

The verdict, as to the beginning of life, is commonly dependent on the standpoint adopted with respect to the possibility of primordial or spontaneous generation (generatio equivoca). This course is, in our opinion, only half correct. The subtlest experiments on spontaneous generation, whether from organic matter or from constituents not yet combined into molecules of organic matter, have proved indecisive, on both sides. Neither the impossibility nor the possibility can be experimentally demonstrated ; it always remains open to the sceptic to say, if nothing appears, that the failure of spontaneous generation is due to the conditions of the experiment ; or if anything does make its appearance, that, notwithstanding every precaution, germs made their way into the infusion. Opinion as to continued primordial genesis still taking place, is thus a mere emanation of the general theory of nature held by each individual. To any one who holds open the possibility that, even now, animate may be evolved from inanimate existence, without the mediation of progenitors, the first origin of life in this natural method is at once self-evident. But even if the proof were given, which never can be given, that in the present world spontaneous generation does not occur, the inference would be false that it never did occur. When our planet had reached the phase of development in which the temperature of the surface admitted of the formation of water and the existence of albuminous substances, the quantitative and qualitative conditions of the atmosphere were different from what they now are. A thousand circumstances now beyond our control, and as to the possible nature of which it is needless to speculate, might lead to the production of protoplasm, that primordial organism, from the atoms of its constituents.

Hence the beginning of life at some bygone period is likewise not susceptible of demonstration; but the commencement of animate being at some definite era of development is a logical necessity, and by no means a vulnerable point in the doctrine of Descent.[54]

We have already incidentally mentioned a man who, although not so eminent as Darwin, has the glory of having independently discovered the law of natural selection, and of having, after Darwin had come forward with his fundamental work, supported the theory of selection by a profusion of original observations. This is Alfred Russell Wallace.[55] In a paper, published in 1855, he demonstrated the dependence of the flora and fauna on the geographical position and geological nature of the district of propagation, and the close connection of the species, according to time and habitat, with kindred species previously existing; and in a second work, in the year 1858, on the inclination of varieties to deviate without limit from the original type, we find a disquisition on the importance of the struggle for existence, the consequences of adaptation, the selection of the most useful, and the replacement of the earlier species by the establishment of the more valuable varieties. We shall repeatedly have occasion to draw upon the rich supplies of his researches.

VIII.

Heredity—Reversion—Variability—Adaptation—Results of Use and Disuse of Organs—Differentiation leading to Perfection.

THE two properties of organic being which determine and regulate the relation of the offspring to the progenitors, and which not only assign to individuals their position in the surrounding world, but also help them to attain it, are transmission or heredity, and adaptation.

Heredity is the conservative, adaptation, the progressive principle. Yet all heredity is not directed to immutability, and many cases of adaptation involve morphological and physiological retrogression. For the elucidation of the inherited peculiarities of organisms, we reconstruct their pedigree ; by the characters acquired by adaptation, we test the pliability of organisms in the lapse of time, and trace the ramifications of the pedigree. Groups of organisms, in which the conservative principle predominates, certainly evince their powers of endurance in the struggle for existence, but they make no advance in physiological value, and are outstripped by the more progressive groups which yield to obstacles and profit by them, a course of which human life also affords so many examples.

As the phenomena of heredity are usually more obvious than the results of adaptation, the latter was almost entirely neglected by naturalists in former days. And indeed what comparison in organic nature can be made so frequently and universally as the resemblance of the offspring to the parent? An anatomist, it is true, quaintly attempted to work out the proposition that the resemblance in the children is not dependent on heredity, but is the result of identical and similar influences, customs, and habits, prevalent in families. But this paradoxical theory requires no special refutation. It is quite true that similar habits and similar external impulses elicit a certain similarity of demeanour and appearance; but if the little son of the pompous millionaire apes his father, it cannot be said that he has likewise mimicked his large or small nose, &c., or has acquired it by a similar call for adaptation. We have only cursorily alluded to this quibble, in flagrant contradiction as it is with every experience; and, in conformity with general opinion, we corroborate the transmission of the parental characteristics to the offspring. The breeders of animals in particular has occasion to observe these transmissions specially, and to evolve their astounding progress from the combination and reciprocal influence of the various forms and degrees of heredity.

It is well known that not only are normal conditions transmitted, but monstrosities are also reproduced through several generations, and, as we have seen in the instance of the crook-legged sheep of Massachusetts, may even be established as the characters of a race. A mere reference to the inheritability of morbid tendencies, bodily and mental, will enable us to realize this intrinsic connection of the offspring to the ancestors. Only since the theory of selection has rendered the modalities of the transmission of bodily characters a subject of more profound study, have general and national psychology been impelled to estimate the influence of heredity in the province of the mind, and demonstrate how, in the various races and families of nations, the molecular peculiarities of the brain, the tendency of character and intelligence of the individuals, and whole series of ideas, conform both in vigour and purport to the laws of heredity.

It is manifest that the key to the phenomena of heredity must be looked for in the process of reproduction. The molecular motions and disturbances, the inconceivably minute mechanical transfers which take place, do not, indeed, admit of observation. They are, however, no more "obscure" and "enigmatical," as they are so readily termed, than the invisible, but not supernatural motions, on the control and calculation of which the stately edifice of theoretic Chemistry and Physics securely rests. With the advance from asexual to sexual reproduction, and from the simple to the more perfect organisms, the difficulty of representation increases, but not that of abstract comprehension. If a low organism, a monad, divides itself, the divided individuals differ from the parent individual only in their inferior bulk, and the difference of their functions is, as to quality, nil.

So, too, where gemmules and germs separate from a parent organism, the dower of the offspring is so large that identity in form and function of progenitor and progeny appears self-evident and natural. But the sexual reproduction of composite organisms is, as we have known since the old doctrine of the *aura seminalis* was refuted, also a separation of material portions of the parental organisms. It is still a mechanical process which is not incomprehensible, and seems inexplicable only if we make the naturally futile attempt to bring sensibly before us the infinitely minute agencies which operate both mechanically and chemically. In the "Variation of Plants and Animals," Darwin has set up a provisional hypothesis of Pangenesis. He says that all phenomena of heredity and reversion would thereby be rendered possible, that in every elementary or cellular portion of the organism innumerable gemmules are produced, which are hoarded up in the reproductive elements, in every ovum, in every sperm corpuscule, and might remain latent during hundreds of generations, and only then exhibit their powers in reversion.[56] This hypothesis, it appears, has met with no ready approbation, probably, as it seems to us, because, in the attempt to meditate upon it, the sensible representation forces itself forward only to prove inadequate. But if it be steadfastly borne in mind that in Protoplasm, as Rollet[57] appropriately terms it, the most complex phenomenal forms of life possess a most persistent witness of their connection with the simplest, it follows that the general laws shown to be true or probable with reference to the simplest organisms, must be applicable to the most perfect also. This holds good also in reproduction, which, in its fundamental phenomena, offers nothing that cannot be based upon molecular physics applied to colloidal living substance capable of imbibition, and thus divested of vitalistic dualism.

The more highly complex is an organism, that is, the greater the differentiation in the development from the protoplasm of the germ-cell to maturity, the more heterogeneously does heredity display itself. These modes of heredity have been defined by Darwin, and yet more systematically by Haeckel, as "laws of inheritance," and corroborated by them by a profusion of examples. If heredity may be termed the conservative element in the life of species, we may also speak in particular of a conservative heredity, by which the old, long-established characteristics and peculiarities are transferred. The more stubbornly a character is transmitted, or, what amounts to the same, the greater the number of families, genera, and species, over which a character is extended, the more ancient must it be considered, the earlier did it appear in the ancestral stock. In most cases, this conservative heredity occurs in an unbroken succession of generations, an observation on which it is needless to enlarge, as it may be daily made by every one. But conservative heredity may likewise display itself intermittently, either when merely individual characters of the ancestors reappear after lying dormant for one, several, or many generations,—a phenomenon designated as Atavism, or reversion,—or when the species is composed of a regular alternation of variously constituted generations and individuals. This particular sort of reversion is termed Alternate Generation, or Heterogenesis.

No one is surprised if children exhibit the bodily or mental features of their grand-parents which were suspended in the parents. But most frequent and striking is the atavism of domestic animals and cultivated plants, a stubborn antagonist to breeders. Of no domestic animal is the aboriginal stock known with such approximate certainty as that of the pigeon. Now there are races of pigeons purely bred for several centuries, and in colour and shape transformed into new creatures, which yet from time to time spontaneously, or by crossing with other conspicuous races, produce birds which, in colouring and characteristic pencilings of black bars on wings and tail, resemble the wild rock-pigeon.

"I paired," says Darwin,[58] "a mongrel female barb-fantail with a mongrel male barb-spot, neither of which mongrels had the least blue about them. Let it be remembered that blue barbs are excessively rare; that spots, as has been already stated, were perfectly characterized in the year 1676, and breed perfectly true; this likewise is the case with white fantails, so much so that I have never heard of white fantails showing any other colour. Nevertheless, the offspring from the above two mongrels was of exactly the same blue tint over the whole back and wings as that of the wild rock-pigeon of the Shetland Islands; the double black wing-bars were equally conspicuous; the tail was exactly alike in all its characters, and the croup was pure white."

Another reversion frequently to be observed is the striping of the feral domestic cat of Europe, in which it resembles the wild-cat so closely as to be scarcely distinguishable. Darwin has collected evidence from which we may infer that the wild ancestral stock of the horse was striped, and this evidence includes the appearance of striped individuals. But yet another strange phenomenon in horses may be interpreted by atavism. Foals are occasionally born with supernumerary toes. This "monstrosity" can be explained only by reversion to the three-toed historical ancestors of the present genus. These vouchers are sufficient.

All the phenomena of artificial breeding, as well as natural selection, serve to show that not only the characters descended from past ages, but also those subsequently and most recently acquired, may be transmitted to posterity. This is progressive heredity. Without it, improvement and progress would be impossible; and its own possibility is the direct result of the nature of reproduction. The newer a useful modification, the less has it hitherto been able to place itself in correlation with the entire organism, the less is the reproductive system as yet affected by it; the more uncertain and fluctuating, therefore, is the transfer by propagation; breeding, or natural selection, is requisite to convert the potentiality of progress into a fact, and gradually to enrol this fact among the conservative inheritances.

Progressive heredity is naturally more complex where the sexes are separate, where sexual selection asserts its rights, and the advantages of one sex are fostered by the taste of the other, and are then either transferred exclusively to the sex benefited by its secondary characters, or turned to the profit of the whole species. As a rule, the males are endowed with these advantages, and have transmitted them incompletely to the females. We will explain ourselves by a single example. In the order of insects termed Orthoptera (or straight-winged), the males, by rubbing their wing-covers together, or by stroking them with the lower portion of their hind legs, are able to make a music attractive to the females. Von Graber, a distinguished modern entomologist, has shown[59] that the teeth of the stridulating instruments of these animals are merely modified hairs; that their construction may be explained by their use; and that in all probability they have been perfected by sexual selection, the best and loudest musicians being the most favoured wooers. With one single exception, the females of the Orthoptera are dumb, but many possess traces of the stridulating apparatus peculiar to the males. Contrary to the older opinion, that it was merely a case of transmission emanating from the males, Graber has made it "more than probable that the resonant nervures of the females—of the stridulating Ephippigera vitium—have been gradually developed independently of the males, but in the same manner." In other cases, on the contrary, the feebly developed nervures of the females, unfit to produce audible stridulations, seem to be an inheritance from the males.

Heredity at corresponding periods of life is a well-known phenomenon. The tendency to disease is transmitted from the father or the mother to the child to break out at the age at which they suffered. Generation after generation, the milk teeth make room for the permanent teeth at a corresponding time. But all special cases are mere results of the general law of development, by which in the individual characters appear in the sequence in which they were historically acquired and became susceptible of transmission. Heredity, at a definite age after the period at which we consider actual development to be complete, is after all only a continuation of the embryonic development, beginning with fission, germ and ovum, of which the ninth chapter will teach us the signification. In this development of the individual, or ontogenesis, as will be shown below in more detail, processes are frequently abridged or totally omitted which once, while they were being acquired and after they had been established, occupied a longer period, but in the course of selection either became of less importance to the individual, or preserved a physiological value only as phases of transition.

The second great class of characters, namely, those which have been newly acquired and depend on adaptation, pre-suppose the mutability of the organism. This is a fundamental phenomenon of organic bodies. It is inherent in the minutest morphological constituents, in protoplasm, and in cells, and in the morphological elements evolved from them, the pervading and determining individual life of which results in the collective life of the creature. The organic morphological element is in a state of saturation; it is continually imbibing and emitting, and its stability is therefore constantly dependent on the supply of material for its functions. For nutrition, which generally and wholly determines the external appearance and the nature of the individual, is accomplished by the innumerable cells and their derivatives. Every fluctuation of supply in any part of the organism, nay, in any point in the surface of a microscopic reef-builder, must necessarily involve a modification of textural parts, or of integrated textural groups or organs.

Mutability is thus a character resulting from the intrinsic nature of organism, and dependent on external conditions which determine quantity and form, as well as the development and transformation of the elementary constituents, or their abortion and retrogression. These effects may be exhibited in a polypestem, which as a whole represents the individual, in its single polypes, the cells and morphological elements. The single individuals are alike in their constitution, but are usually very different in size and development, even in those species in which the differentiation unquestionably produced by selection has not led to polymorphism or separation into personal groups performing different functions. The weal or the woe of our polypes is greatly dependent on the position which they occupy upon the stem; the supply of nutriment primarily furnished to the single individuals is unequally and variably apportioned according to currents and tides. Hence on each polype-stem there are regions where the single polypes are especially thriving, others where they are just able to maintain themselves, others where they cannot keep their balance. But as the polype-stem is traversed by a canal system conveying the nutritive fluid and connecting the several cells, the superfluity of the well-situated cells goes to the benefit of those for whom a worse lot was prepared by their accidental position, and conversely. These relations, which, complex as they seem, are very simple for our comparison, determine the form and appearance of the polype-stem. Among a hundred thousand stems, no two will be found absolutely alike.

To return to the mutability of organisms, even if two individuals of the same species are bred under the most similar conditions imaginable, it has never been possible to pronounce them absolutely alike. That mutability is slighter in lower than in higher organisms, is a prejudice frequently repeated and fortified by the old dogma of species. The doctrine of descent and selection would fare ill if the case were so. But as the shepherd unerringly knows the physiognomy of his sheep where an excursionist from the town sees only a general sheep's face, so to an attentive naturalist, in most of the lower organisms, the specific type resolves itself into as many varieties as individuals, irrespectively of the cases in which no specific type can be established.

As modification under given conditions, adaptation is thus as little an unknown quantity as heredity, but is merely a function of the mechanical character of mutability, or, in the widest sense of the word, of nutrition. Adaptation takes place when the organism or its parts are pliable and plastic to external influences, when they conquer and make use of them. Climate, light, humidity, nutriment, are hindrances or advantages that directly or indirectly affect the organism, and are all actively concerned in it. Surrounded by organisms, we see them without exception adapting themselves to circumstances; and if our only object is to be convinced of the formative influence of the mode of life, this is most readily done in the case of domestic animals. In his studies on the pig, H. von Nathusius, perhaps the most scientific of the celebrated breeders, shows how in the simplest cases, where the looseness of cultivated soil has facilitated the labour of grubbing, the skull of the domestic pig is arrested, by the softer structure of the cranium, at the immature form of the wild boar, and how those extreme shapes of the head in cultivated breeds, characterized by the bending and shortening of the face, and the impossibility of closing the jaw in front, are entirely the result of their altered mode of life. It is known that men, animals, and plants, removed far from their previous abode to a new and strange environment, after a longer or shorter effort of the organism to domesticate itself, either die out, or else accommodate themselves to the new conditions and become acclimatized. Every acclimatization is therefore an adaptation, accompanied by modifications more or less perceptible. Thus, in consequence of the varied conditions of life, there is a wide divergence among races of men who, by their kindred language, are of the same origin, not to mention those whose relations linguistic inquiry has not yet decided. How different is the idiosyncrasy of the Englishman from that of the Hindoo! Physically and psychically, they represent two remarkable sub-races of which the peculiarities must be ascribed to adaptation,—in the latter, to a climate which requires a vegetable diet, and, eliciting neither bodily nor mental energy, favours a dreamy sensuality; in the former, to a country which is in every particular the opposite of the Indian original home. Similarly, the annual alternation in the vital phenomena of so many organisms, designated as hybernating animals, is a case of adaptation. It is changed the moment the organism is exposed to another climate, or rather acclimatization is essentially the accommodation of the hybernating animals to the new climate.

In all these examples we have the results of direct adaptation, in which the power of resistance in the individual comes into play, as does cumulative adaptation in artificial, and the survival of the fittest in natural, selection. In all cases of adaptation, one or several organs are primarily concerned, either actively or passively; and only in consequence of the resulting modifications are the other organs drawn into sympathy. This may be termed correlative adaptation. It might be supposed that the most perspicuous examples would be afforded by parasitic animals, in which, with the alteration of the aliment and of the alimentary apparatus, especially of the manducatory portions, is usually combined a transformation and retrogression, often extending to the total extinction of the locomotive organs, and of the entire segmentation of the body. But, although the limits are difficult to define, the cause of these associated modifications in the locomotive and alimentary apparatus consists less in their reciprocal sympathetic influence than in their simultaneous disuse.

It is, however, by correlative adaptation that, for instance, in the short-beaked races of pigeons, the middle toe and astragalus are shortened, and that in the long-beaked races these organs have shared in the elongation. In the case, however, in which short beaks are combined with short feet, a certain share in the shortening of the feet is also owing to disuse; while where the pigeon-fancier took pleasure in the elongation of the beak by cumulative selection, the correlative elongation of the foot took place in spite of disuse. The most important group of correlative modifications or adaptations, always using this word in its widest acceptation, relates to the sphere of the sexes. Direct attacks on the generative organs manifest their effects on all the rest of the organism, as is best shown in animals of both sexes castrated for the market or for labour.

We have already seen that the degree of perfection attained in the orders of the Articulata, Annulosa Vertebrata, and partially in the Radiata also, depends on the integration of the originally similar parts lying behind or by the side of one another; hence on the division of labour. This Haeckel has designated divergent adaptation. It gives rise to the remarkable polymorphism, which appears especially in the marvellous forms of the Hydra tuba; and, higher up, in the segmentation of the classes of the Termites and the Bee, &c.

So far as modification coincides with adaptation, the direct adaptations hitherto discussed may be opposed by a series of indirect adaptations. Among these may be comprised a series of phenomena of which the causes do not fall within the life of the individual, but are to be sought in influences by which the parents were affected. It is obvious that we here come into contact with the province of heredity in a manner well known to breeders. Thus H. v. Nathusius, in his studies on the formation of the pig's skull,[61] says:—"From the facts here collected, it is plain that the transmission, the transfer of the form of head from the parents to the offspring, does not unconditionally ensue. If the form of skull, which we will briefly term the cultivated form, be the product of nutrition and mode of life, hence of external influence,—if it can be differently formed in the same individual, and is therefore not constant,—in that case the heredity of this form cannot be spoken of without qualification. The form itself will not be transmitted to the offspring, but only the tendency to the form. This may be inferred from the circumstance that from generation to generation, and to a certain degree, the form increases in peculiarity. If we rear a common with a thoroughbred pig, and if we allow exactly the same influences of nutriment and keeping to operate upon both, and in equal measure, we shall not obtain the same form of head. The development of the form of head must therefore be aided by a pre-existing tendency, and we must hence regard it as hereditary."

Haeckel likewise propounds a law of individual adaptation, which expresses the fact that, notwithstanding the closest kinship, individuals diverge in many ways. The cause of this difference, chiefly conspicuous in the individuals of the same litter or brood, is, so far as it is not due to adaptation, inherent in the germs, and is transferred to them by fluctuations and differentiations in the conditions of nutrition in the parents, mostly beyond our ken. Other phenomena of indirect adaptation are exhibited in the occurrence of malformations, of which the causes must be looked for only in disturbances of nutrition in the parental organisms by which the progenitors themselves were not perceptibly affected. Here also belong the cases in which influences which have affected one sex only are manifested exclusively in posterity in the same sex. As may be seen, these processes, of which the initiation is entirely withdrawn from observation, are closely connected with the most obscure province of heredity.

An extremely interesting and important form of adaptation is the so-called mimicry, or protection by means of colouring and form. The first discoveries on this subject were made by Bates, the well-known "Naturalist on the Amazon;" the greater part were subsequently added by Wallace. In South America, the family of butterflies named Heliconida is extraordinarily extensive; they are remarkable for their elongated wings, body, and antennæ, and for the beauty of their colours. It might be imagined they were exposed to the persecutions of insectivorous birds and other animals ; but this is not the case, for they have a disagreeable smell, which, in all likelihood, renders them obnoxious. Their smell and flavour are thus a protection, as the birds and lizards who have once seized them by mistake are certain, ever after, to leave them unmolested. Now, as the insectivora do not test the individual case, but have adopted a general repugnance to the aspect of the Heliconidæ, if other butterflies resembled the Heliconidæ without possessing the bad smell, they would participate in the security to life enjoyed by the Heliconidæ in proportion as they approach their external appearance. This case has actually occurred, for Bates discovered a number of species of the otherwise very different genus, Leptalis, of which each almost undistinguishably resembles one of the Heliconidæ both in colour and form. The Leptalidæ have also adopted the flight of the Heliconidæ, share their habitats, and, although without the offensive smell, fly about with impunity. This state of things would be impossible if the Leptalidæ were not considerably in the minority, so as to be in a measure hidden by the Heliconidæ.

Wallace has proved that species protected by mimicry of other animals are invariably in the minority, and often very rare in comparison with the species which they imitate. Neither the explanation that like conditions of life produced like results, nor the hypothesis that, in some cases at least, the mimicry consists in reversion to a common original species, is in any way satisfactory. Many cases can be interpreted only by natural selection, those, namely, where from the first, before the imitation had begun, such a resemblance already existed between the imitating and imitated forms as to render confusion possible; where, therefore, the resemblance so conducive to the preservation of those in which it was the strongest, needed only to be increased by natural selection. Darwin[62] is also of opinion "that the process probably has never commenced with forms widely dissimilar in colour."

A peculiar, simpler and long known mimicry, is when animals have accommodated themselves in colour to their habitats in such a manner as not to attract the attention of their enemies, and likewise to deceive their prey. Who, in the days when he chased butterflies, did not learn how difficult it is to recognize certain evening and nocturnal flyers on the bark of trees, as they quietly sit with their dusky brown or gray-striped or speckled wings, outspread in a roof-like shape? The tree locusts and Mantidæ can look so deceptively like leaves or twigs, that it is only by the touch that one can be assured of their real nature. Wallace relates that one of the Phasmidæ (Ceroxylus laceratus), which he obtained at Borneo, was so covered with pale olive-green excrescences, that it looked like a stick covered with moss. The Dyak who brought him the animal declared that, although alive, it was really overgrown with moss, and the naturalist himself was only convinced of the contrary by the closest examination.

A remarkable example of advantageous colouring, within easy reach of many of our readers, is exhibited in most species of the flat-fish (Pleuronectidæ), now so frequently kept in aquaria. Observe the gray or brownish speckled creatures, as with a few strokes of their fins they partially cover their upper surface with sand. They need not bury themselves entirely, for it is only by close examination that their bare skin can be distinguished from the sandy bottom, and under this partly artificial, partly natural veil and mask, the animal waits for its prey.

In many animals provided with protective colouring the phenomena are more complex, and explanation by natural selection is far more difficult; for they are able voluntarily to adapt their colour to circumstances, or else their colour changes by involuntary reflexes. Verany's unsurpassable observations on the Cephalopoda have acquainted us with the range of colours at the disposal of these Molluscs, and to this may be be joined Brehm's description of the changes of colour in the chameleon. On these highly complex cases some light is thrown by the simpler instances in which the manifestly protective colouring has become fixed in skin and plumage, and the concurrence of other circumstances scarcely admits of any other explanation than selection.

On this point, Wallace's interesting researches on bird's-nests are especially instructive. The great majority of female birds which sit in open nests possess brown or gray, in short, unobtrusive plumage. No contradiction will be offered to the statement that any casual modifications of plumage, which would more readily betray the sitting bird to its enemies, would have no prospect of becoming constant. The converse follows naturally with regard to colouring which brings the bird into harmony with its environment; and an important guarantee of the correctness of this interpretation of facts is afforded by the other observation, that most female birds with gaily coloured or speckled plumage sit in covered and concealed nests. It must be added, that the construction of nests is not determined by the absolute rules of a blind instinct, but is modified by the experience of the animals, an experience of which we are indeed scarcely able to perceive the development, except with the age of the individual, but which, at least in several cases, has been proved to be the progress of the species.

Natural selection has an important accessory in the modifications produced by the use or disuse of organs. Compulsion to more diligent use, inducements to disuse, are involved in the varying conditions of life. In both cases it is therefore a question of adaptation. Looking at nature, profound modifications are most readily demonstrated as the consequence of disuse; but artificial selection gives numerous examples of both sorts, especially where disproportionate use of certain organs is combined with simultaneous disuse of others. Such products of selection with disproportionate use are the racer and the dray-horse.

The blindness of cave animals admits of no explanation, but that, with the increasing uselessness of the eyes during accommodation to cave life, the exchange of material in the less active organs gradually diminished, and atrophy was initiated. The accuracy of these theoretical observations is enforced by the observation that the nearest kin of many blind cave animals, especially of insects and spiders, reside in the vicinity of the cave, and that those cave animals which inhabit passages only partially obscure, possess less atrophied optic apparatus. A singular gradation occurs among the burrowing mammals, and Darwin[63] cites an example admirably illustrating the loss of sight in consequence of the mode of life. "In South America a burrowing rodent, the Tuco-tuco, or Ctenomys, is even more subterranean in its habits than the mole; and I was assured by a Spaniard, who had often caught them, that they were frequently blind; one which I kept alive was certainly in this condition, the cause, as appeared on dissection, having been the inflammation of the nictilating membrane. As frequent inflammation of the eyes must be injurious to any animal, and as eyes are certainly not necessary to animals having subterranean habits, a reduction in their size, with the adhesion of the eyelids and growth of fur over them, might in such case be an advantage; and, if so, natural selection would constantly aid the effects of disuse."

In the classes of flying animals, a large number have left off flying; and we find their flying apparatus in an aborted or incomplete condition, which perverse judgment and reasoning alone can regard as a state of progressive development from yet simpler rudiments. If throughout the great family of the Coleoptera, genera and species are to be found with imperfect flying apparatus, consolidated wing covers, &c., if the whole family of Staphylinæ does not possess the power of flight, no one dreams of considering them as arrested forms; but it is conceivable that the mode of life in which they differ from the other members of their order and class, gradually superinduced in their flying ancestry the habit of not flying, and at the same time the atrophy of the organs of flight. With this was combined, as these beetles show, no degradation of organization, but, on the contrary, a higher and extremely advantageous development of other organs, the manducatory and locomotive apparatus. A general reduction of the power of flight has been shown in the beetle fauna of many islands. Thus in Madeira, of 550 species, over 200 fly imperfectly or not at all, and for this there is no explanation but natural selection. Here the less good and enterprising flyers had the advantage, while the others were blown into the sea and eliminated. The non-application of a previously attained special perfection is advantageous in the "struggle for existence."

In several families of lizards, some genera are serpentine, as they are termed, which, with elongated bodies, possess either fore-legs only (Chirotes), or merely rudimentary hind-legs (Pseudopus), or no vestiges of legs (Anguis). They bear the same relation to the great class of normally four-legged lizards as the non-flying insects to their own class. They have not been arrested in their development, nor are they animals in process of evolving four legs; but, as Fürbringer has demonstrated from the history of development and comparative anatomy, their limbs, and—if these are entirely absent—the remains of the pectoral and pelvic arches and the sternum bear indubitable marks of the abortion of a once complete apparatus. Further comparison shows that this atrophy reaches its climax in the snakes, but that it is compensated for by the ribs and intercostal muscles having undertaken the work of the limbs. Here, again, disuse and adaptation coincide as well as differentiation.

In the class of birds is repeated the spectacle we have just witnessed in beetles and reptiles. In some few families and smaller groups, individual species are deprived of the power of flight, and one whole large systematic group is characterized by the incapacity of flying. In our opinion, there was a direct connection between the inducements to disuse and its consequences in the case of the dodo, which, with its few congeners, so promptly fell a sacrifice to its helplessness on the discovery of the lonely islands which they had probably inhabited for thousands of years without disturbance. In no other way has the northern penguin (Alca impennis) at some time obtained the curtailment of its wings; and the scanty but wide-spread remains of the order of flightless birds indicate a period at which, in a more peaceful environment, their far more numerous wingless ancestry made less use of their pinions, and natural selection endowed them with greater strength and nimbleness of leg. The effects of disuse of the organs of locomotion are likewise directly exhibited by artificial selection.

Use and disuse, combined with selection, elucidate the separation of the sexes, and the existence, otherwise totally incomprehensible, of rudimentary sexual organs. In the Vertebrata especially, each sex possesses such distinct traces of the reproductive apparatus characteristic of the other, that even antiquity assumed hermaphroditism as a natural primæval condition of mankind. The technical proofs of the homologies concerning these partly manifest, partly internal and hidden relations, are given in the manuals of comparative anatomy. We shall merely indicate the manner in which the theory of selection is here borne out. It is self-evident that in hermaphrodite animals, fluctuations in the sexual sphere must take place, in which one half or the other will predominate. Should these fluctuations be sufficiently strong for natural selection to take possession of them, the productive power of the less active portion will gradually decrease, and finally, with the extinction of the physiological character and the function, nothing will be transmitted but the morphological remains, as a mockery to the theory of special design or teleology. Here and there only occurs a reversion more or less striking, connected, however, almost exclusively with the adjunctive organs, and the secondary sexual characters, by which we mean, not those acquired by either sex, but originally common to both. The tenacity with which these rudiments of sexual organs are inherited is very remarkable. In the class of mammals actual hermaphroditism is unheard of, although through the whole period of their development they drag along with them these residues, borne by their unknown ancestry no one can say how long.

Unless we suppose that parasitic animals were created simultaneously with their hosts from the dust of the earth,—man and his tapeworm, and other disagreeable guests,—and thus put an end to the discussion, this entire province has to be explained by descent, with the special co-operation of disuse. The proposition to be demonstrated in the next chapter, that the evolutionary history of the individual represents the history of the species, will show the influence of the disuse of particular organs on the configuration of the various parasites. The parasitic Crustacea are perhaps the most instructive, as they present the most complete systematic series, exhibiting the gradual atrophy of the organs which accompanies the ever-increasing connection of the parasite with his host. In several orders of intestinal worms, the alimentary canal has become entirely unnecessary; but they exhibit neither intermediate forms nor phases of development. It is different, however, with the parasitic Crustacean, for here the young, locomotive, and well-integrated being has its prototype in definitive generic forms permanently locomotive, from which, after adhesion, it deteriorates into a mere motionless sac. All these animals, including the intestinal worms, have acquired their position and status (and this is the true significance of parasitic life) by the apparent degradation of their organization. They are, almost without exception, distinguished by their reproductive power; and on this, owing to the easy supply of nutriment, without any exertion of the other parts of the organic system, the whole bodily activity could be concentrated.

We have hitherto demonstrated that organisms are urged to continual differentiation by the unremitting struggle for existence. For the cultivation of morphological species, natural selection, moreover, seizes on the modifications arising from the mere variability of the organism, and implying no physiological advance. But sooner or later these are also inevitably drawn into the vortex of competition. After what has been already said, this fact is so self-evident as to need no further proof. Even did we not see the infinite variety of organisms, a divergence into novelty must needs be inferred on *à priori* grounds, from the existence of the simple and uniform, and the necessity of adaptation to altered external conditions. But with development in various directions, under the guidance of natural selection, progress is necessarily combined. It is one of the greatest services rendered by the theory of selection, that it has finally broken with the notion of design, which hitherto invested the organic world with perfection externally bestowed, and even in the province of intelligence and morality, where it is said with Schiller,

> So grows the Man as grow his greater aims,*

has secured admittance for the uniform method of natural science.

It is highly remarkable how the teleological view of nature could be so long upheld, and is still in part upheld, by theological influence although in the whole organic world we behold a merely relative perfection, and the manifest and multifarious arrangements adverse to design in every grade of organisms, bear a bad testimony to the external directing Power. The perfection exhibited by comparative anatomy, and the estimate of physiological functions is, under all circumstances, the result of adaptation and selection. In the struggle of all against all, those individuals win who in any degree excel their fellows in the division of labour, which, if the direction of activity be altered, often obliges them to disuse organs which were once of service, but in the new conditions are useless, and, it may be generally said, have become injurious.

* Es wächst der Mensch mit seinen grössern Zwecken.

Artificial selection—and here we may speak of design—produces perfection when, by mechanical and physiological labour (the latter especially by means of suitable nutriment), it exercises the particular parts which are to be perfected, and propagates the advantages obtained. What we term natural selection is the epitome of the improvements acquired by specialization in the process of adaptation. The most faithful image of this gradually acquired specialization is afforded by the development of the individual, where from the undifferentiated, by constantly increasing differentiation, the mature animal is evolved in the plenitude of its physiological functions. That in the various animal groups certain grades of perfection are attained, is an uncontroverted fact; but every closer investigation shatters the idol of design. The organism of the bird might induce us to consider it, in the abstract, as modified for the purpose of flight. But if design be allowed to watch over the good flyers, the idea of design must be abandoned with respect to the non-flyers, and, if some idea is indispensable, adaptation must have its due. Herewith the whole theory is broken down, and it will be the same in every other case.

How organic perfection stands with reference to the idea of design, has been acutely and clearly expressed by the author of the "Unconscious" ("Unbewussten"). The theory of descent teaches that there is no independence of the conditions co-operating in an organic phenomenon; rather that its increasing divergence from a common neutral point was an effect of the same causes. The theory of selection makes us acquainted with one of these causes, and unquestionably the most important as one, which, by purely mechanical compensative phenomena, produces advantageous results. The theory of descent merely casts doubts on the teleological principle by withdrawing the basis for positive proof, but the doctrine of selection sets it directly aside, so far as it is able to extend its explanation. For natural selection in the struggle for existence, the extermination of the less appropriate, and the survival and perpetuation of the fittest and most appropriate, is a process of mechanical causality of which the steady conformity to law is nowhere infringed by any teleological controlling metaphysical principle. This, however, produces a result essentially corresponding to design; that is to say, it naturally bestows on organisms the highest capacity for life under given circumstances. Natural selection solves the apparently insoluble problem of explaining fitness as a result, without calling in the aid of design as a principle.

In each family—for, as we have seen, what zoologists once designated type, has in the doctrine of Descent become the family—in each family lies the potentiality of a certain grade of perfection; and when the main outline of the family character is established, we see a development taking place, of which the potentiality is inherent in the tendency of the character, the realization and necessity in the external conditions. Hence to us also, progress is development, but not towards a predestined and pre-established harmony. Karl Ernst v. Baer,[64] anxious to rescue design, or at least the "purpose"—in short, predestiny, in the evolutionary series of Nature, says: "Every cause engenders a process which again works on towards another purpose." But why purpose? Ought it not rather to be: Every cause engenders a process which again works on towards another process? The further we go back, the deeper and more general is the grade, and the various ramifications at their peripheral ends have either halted, or arrived at very different grades.

An objection frequently made against this result of the doctrine of Descent is, that if all are pressing forward towards perfection, how is it that, besides the higher, so many lower members of the family are able to maintain themselves, and how can the lower families hold their own against the higher, in the struggle for existence? In presence of the irrefutable facts of progress, it is enough to point out that the lower forms could and can continue to exist wherever they could find space as well as the other necessaries of life. While they here underwent only slight modifications, elsewhere the needful selection led to more profound metamorphoses; and on a subsequent geographical displacement, the newly transformed beings, accustomed to other conditions of existence, were again able to share sea and land with the stationary species. For as diversity is restored now by selection, and the demands for nutriment and other necessaries are likewise different, a partial remission in the struggle must take place.

The preservation of a great many inferior organisms is evidently favoured by the circumstance that just because they are simpler, their propagation is more easily effected. Hence although, especially in limited districts, amid violent competition of superior varieties, countless species must suffer extirpation, yet the struggle for existence and perfection do not exclude the existence of lower forms. But teleology, as it seems to us, still owes an explanation of what has been explained by the theory of selection. The retardation of the lower organisms, notwithstanding the internal pressure and the appointed purpose, is incomprehensible.

But, it is frequently asked, if you will not hear of a "principle of perfectibility" inherent in organisms (Nageli), of the "divine breath as the inward impulse in the evolutionary history of nature" (Braun), of "tendency to perfectibility" implanted by the Creator (R. Owen), even of the "striving towards the purpose" (v. Baer), can chance be supposed to have produced these marvellous higher organizations? To this it may be plainly answered, that this chance, to which purblind humanity allots so great a part wherever the personal interference of a superior Being or the universal "creative and productive principle" is not at hand, has no existence in nature, and that our conviction of the truth of the doctrine of derivation is due to its adjustment of the phenomenal series as causes and effects. Let us remember, and fancy ourselves in possession of, the formula of the universe of Laplace, by the aid of which all future evolutions might be computed in advance. With our limited powers, it is true, it is retrospectively alone that certainty can be approached in the calculation and discrimination of the series. In this we must obliterate the word chance, for causality, as we understand it, makes chance entirely superfluous. Any one who transports himself to the commencement of an evolution, who, for instance, fancies himself present at the genesis of the reptiles, may, from his antediluvian observatory, look upon the development of the reptile into the bird as a "chance," if he does not peradventure regard it as predestined. To us, who trace the bird backwards to its origin, it seems the result of mechanical causes.

Let us now recapitulate what we have gained by the doctrine of Descent, based on the theory of selection; it is the knowledge of the connection of organisms as consanguineous beings. The greater the accordance of internal and external characteristics, the closer is the kinship. The further we trace the pedigree to its origin, the fewer become the characters persisting to these roots, the more do these characters reveal themselves as acquisitions in the lapse of time. As we eliminate these acquisitions and the inherited characters, the further we probe, the more do we restrict and reconstruct the pedigrees of the various groups.[65]

We do the very thing which in linguistic inquiry is deemed extremely natural and scientific. The ideas and words common to the individuals of a linguistic family are the inheritance from the intellectual and linguistic property of the original people, from which the pedigree of the family has ramified. This so-called "chance" prevailed in the formation of the derived languages neither more or less than in the evolution of organisms from their original forms.

IX.

The Development of the Individual (Ontogenesis) is a Repetition of the Historical Development of the Family (Phylogenesis).

ALTHOUGH the palæontological record is full of gaps, it is nevertheless unmistakable, as even most of the opponents of the doctrine of Descent are ready to admit, that from the older to the more recent period, a progress takes place from the lower to the higher grades of organisms, which is likewise exhibited in the system of the present vegetal and animal world; and that in many ways embryonic development as well as metamorphosis and heterogenesis,—in a word, individual development ("Ontogenesis," Haeckel) suggests a comparison with these palæontological series, as well as with the systematic order of succession. The parallelism of the palæontological and the systematic series is either a miracle, or it may be accounted for by the doctrine of Descent. There is no other alternative. And the doctrine of Descent fully bears the test; it shows how the derivation of the present organisms from those previously existing rests on the transmission of the characters of the progenitors to the offspring and the acquisitions of the individuals. The phenomena of individual development or Ontogenesis admit of no other choice; either they remain uncomprehensible, or they stand the test of the doctrine of Descent and submit to the great general principle.

If we scrutinize the countless facts of reproduction and development, they certainly admit of classification; they range themselves in analogous and homologous groups; types of development become apparent; we speak of development without metamorphosis, of transformation, and heterogenesis. But what necessary relation the alternating forms, the shapes appearing in heterogenesis, bear to the complete animal or the sexually developed chief representative of the species? why so many animals undergo no transformations, but emerge "complete" from the egg? why the species belonging to the same class or "type" possess the same type of development and process of construction?—these and similar questions as to the interpretation of the tangled mass of facts press themselves upon us. And they are also tests of our theory of derivation. The doctrine does as much as has been done by any great hypothesis in its special application; and if it gives a satisfactory reply to all, or at least to nearly all, pertinent questions, these are so many witnesses and proofs of its truth, which, according to all scientific custom and justice and philosophic method, will remain valid until the falsity of the inductions and inferences has been demonstrated and a better hypothesis substituted in its stead.

The first proposition derived from the doctrine of Descent in explanation of the facts of individual development may run thus: accordance in the outlines of development is based on similar derivation; or, somewhat differently stated: accordance in the outlines of individual development is accounted for by similarity of derivation. As we already know, C. E. v. Baer first demonstrated that the members of the great divisions of the animal kingdom agreeing in the outlines of their organization testify their coherence by a special "type of development." This fact was always looked upon as self-evident, although, if it were not derived from descent, it would be the greatest miracle. This is therefore the place for us to review some of the fundamental forms of development which we partially considered in the third chapter, and at the same time to elucidate the meaning of these types with the aid of the doctrine of derivation.

We will take the Echinoderm as our first example. Although from the anatomical comparison of a crinoid, a star-fish, a sea-urchin, and a sea-cucumber or holothuria, the close kindred of these various divisions of echinoderms is easily deduced, they yet deviate wonderfully from one another in outward shape and in the construction of the skeleton. The relative value of the difference between a holothuria and a star-fish, a sea-urchin and a comatula, may be compared to the difference between a mammal and a bird, an amphibian and a fish. Nevertheless, with some few exceptions which have a special meaning, these various echinoderms leave the egg in a larval state almost identical. The larva (Fig. 12) is boat-like in form, with a curved margin bent over at both ends like a deck. This border is edged with a continuous row of cilia, by the agency of which the little boat is moved. A short digestive canal, provided with a gastric enlargement, is the first essential organ of this

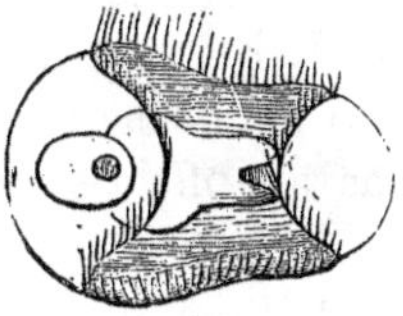

FIG. 12.

body. We will not describe the highly complex transformations of the larva here into an ophiura, there into a sea-urchin, and here again into a sea-cucumber ; but we will only inquire what can be the cause of this accordance in the earliest stages of individual development. There is no reasonable answer but the derivation of all echinoderms known to us from an older form, in the development of which our larva likewise appeared, and from which this common phase of development was transmitted to the whole family. But it is allowable to ask, further, how from a bilateral larva; one, that is, symmetric on two sides, should be evolved in animals of radiate structure, as are the greater number of mature echinoderms ?

On this point Haeckel instituted a conjecture which at first exasperated the systematizers of the old school, but which now gains more and more footing, and is supported by the most recent comparative investigations, such as those of Hoffmann "On the Minute Anatomy of the Starfish" ("Ueber die feinere Anatomie der See-sterne"). The boat-shaped larva of the Echinoderms, especially a modification occurring in the starfish, strikingly resembles a certain larval type of the marine Annelida. And as in the structure and distribution of the parts of the rays of the echinoderms, especially of the star-fish, an unmistakable resemblance with the relative distribution and succession of parts of the Annelids is observable, Haeckel regards the Echinoderms as an offshoot of the Annelids. He considers that the oldest, and to us unknown, echinoderms originated as annelid stems ; the anterior end of the bilateral annulose parent-animal budding out gemmules in a radiate arrangement. This gemmation, or, in other words, this stem structure, still occurs in Echinoderms, inasmuch as some species of star-fish possess such powers of reproduction as to enable a single arm or ray, when torn off, to complete itself into a whole animal. Nay, Kowalewsky's observations render it highly probable that the separation of rays, and their completion by gemmation, is in some species a normal process. Haeckel's hypothesis is thus laughed at only by those who are afraid to think or reason.

In the family of the Mollusca, the so-called navicula larva testifies the kinship of at least two of the great classes. The third and most advanced class, that of the cuttle-fish, had perhaps lost their distinctive badge even in those primæval times when, under the somewhat lower forms of the Tetrabranchiata, they left their shells in the Silurian strata. But the bivalve shells, or Lamellibranchiata, and the snails, widely differing in anatomical development, and constituting two natural classes, have a common larval form, or, if the larvæ display different shapes, a highly distinctive common larval organ, the velum. The accompanying diagram gives on the right the navicula of a cockle-shell as seen from behind. At the anterior end, two fleshy lobes have been formed, edged with cilia, by the vibrations of which the young animal, even in the egg, performs spiral twisting motions; in the midst of the cilia rises a little prominence, furnished with a longer filament. These ciliated lobes or vela, merging into one another, are shown on the left in the larva of a sea-snail (Pterotrachea), as seen nearly in profile, and in the phase in which the eyes and auditory apparatus, the foot and operculum, as well as a delicate shell, have made their appearance. Here also, from the plane of the velum, a small fleshy protuberance juts out, without any special purport. The distribution of the velum, the period at which this larval organ makes its

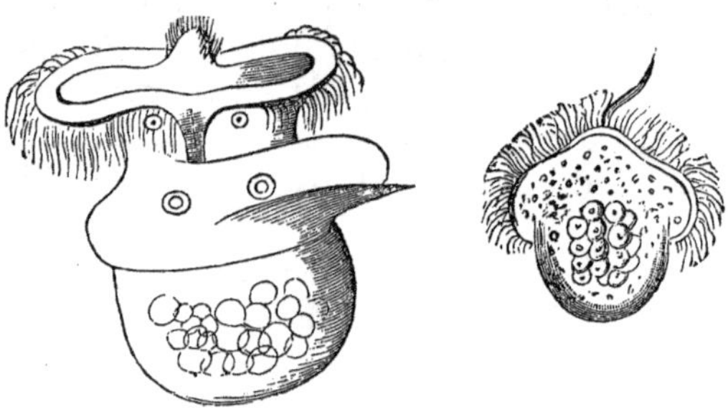

FIG. 13.

appearance, its position towards the testa, head, mouth, and foot, and its subsequent effacement, one and all coincide exactly in the two classes. It is as yet of only a relatively small number of marine shells and slugs that we know the evolutionary history; yet we may infer that in these animals remaining in their original home, this heirloom has been generally preserved. Even genera which in their mature state scarcely recall the type of the Mollusca, as the boring mollusks (Dentalium Teredo), have preserved the phase of the navicula. On the other hand, in the branchiate fresh-water snails (Paludina) the velum is little developed, and in the land snails, which differ most widely from their marine kindred, the velum is entirely obliterated, as it is also among fresh-water mussels. If in these animals adaptation and migration to land has had this effect on embryonic and post-embryonic development, we must suppose that in the Cephalopoda, notwithstanding their continued sojourn in salt water, other causes have produced the loss of the velum phase, and the course of development peculiar to it.

With respect to the other fundamental forms of development, we may refer to the third chapter. The construction of the higher Articulata points to annulose progenitors, more or less corresponding to the annelids of present times; and, again, the gradual increase of the segments of the larval annelids, which may be compared to the process of gemmation, leads from these higher Vermes to the lower ones with unsegmented bodies. All vertebrate animals, man included, if they do not preserve through life an unsegmented vertebral column, not separable into single vertebræ, are raised as embryos from this condition into their higher and definitive phase. That they should pass through this common embryonic condition is prohibited by all other mechanical causes but that of a common derivation from primordial forms which possessed an unsegmented vertebral column, no cranium or an imperfect one, and either no brain or one little differentiated from the spinal cord. Karl Ernst v. Baer, who, while we write these pages, raises his voice against the doctrine of Descent, has established the fact of types of development, and the course, within these types, from the undifferentiated to the special; but by the words "type of development," the fact is paraphrased, not explained; and, as we cannot repeat too often, we prefer the distinct idea of derivation to the supposition of an unknown higher Power manifesting itself after an incomprehensible fashion in the types of development.

If the concatenation of the series by direct derivation and heredity be disallowed, it is absolutely inconceivable why the supreme creative Power, Nature, or the personal God, should have bound all higher animals to the same common stages of early development, and hereby exposed them to such manifold purposeless arrangements and great dangers. Of the millions of young oysters which annually escape from the egg, the majority perish under the disadvantages of external conditions, because the oyster has not yet divested itself of the ancient heirloom of the roving navicula. It has been able to compete successfully in the struggle for existence, only because, like most of its congeners, it is enormously prolific. This may be understood ; but that a personal Creator, merely on principle, in order to keep the oyster within the type of development, should have endowed it with the phase of the navicula, in this case so extremely unpractical, can be accepted, like much other nonsense, only as matter of faith.

If accordance in the outlines of development has generally shown itself derivable from similarity of descent, we may now proceed to the explanation of those phenomena of development known to us as heterogenesis and metamorphosis. In these, the historical stages of development of whole classes and orders are inherited in the development of the individual; a proposition which is merely the corollary and application of what has been already intimated. In no class is there such a profusion of the phenomena of heterogenesis, readily submitting to explanation, as in that of the Medusæ. We have already (p. 43) become acquainted with the origin of the Cladonema from the polype-like

Stauridium. The Medusa is the sexually mature form of the cycle of the species; its ova develope into polypes, which constitute the intermediate form in their development; that is to say, it is not transformed into the

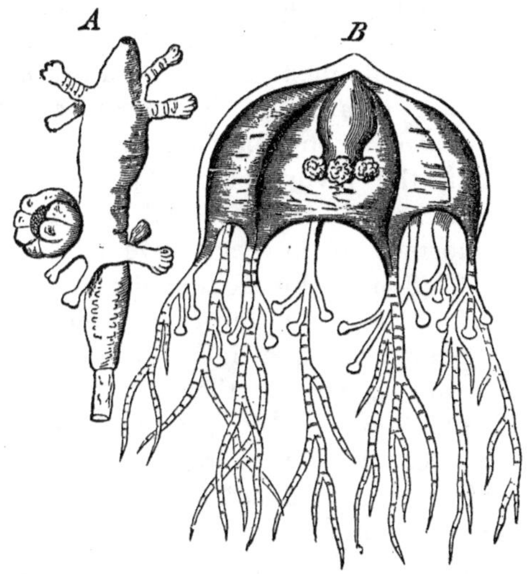

FIG. 14.

animal from which it is derived, but produces gemmules. Only in this generation does the species revert to the sexual form.

We shall understand this alternation of generations if we begin with the simplest Medusa polypes. Such a one is the annexed Hydractinea carnea, of which the female individual is portrayed. Compared with the intermediate form, Stauridium, the preliminary phase of the Cladonema, reproducing itself asexually, the Hydractinea seems superior, inasmuch as it is itself a sexual form. The zone of spherical protuberances in the middle of the body are the ovaries or egg capsules corresponding to the sperm capsules of the male individual. Heterogenesis does not take place in our Hydractinia, but, as in the development of the ovum of the Cladonema into the Stauridium, there is a transformation of a ciliated lava into a sessile polype. But it is obvious that the part which in the Hydractinia is played by the male and female sexual organs is performed in the generative cycle of the Cladonema by the sexual animals. By following the transition from the dependent organ into the independent animal, we find the solution and explanation of the process termed heterogenesis. Between the genera reproduced like the Hydractinia, and those reproduced like the Cladonema, there are many others, of which the propagation shows the gradual transition of the rudimentary sexual organs into the sexual animal. We may so arrange the genera of the "Medusa polypes" as to exhibit how the parts which in the Hydractinia are mere capsules, generating and enclosing the ova, become more and more perfect. They acquire a special branch of the alimentary canal and blood-vessels, and are provided with the marginal papillæ characteristic of the Medusæ, and constituting their peculiar sensory organs. In short, what in one member of the systematic series may be termed an organ, is, in the next, the Medusa separating itself and

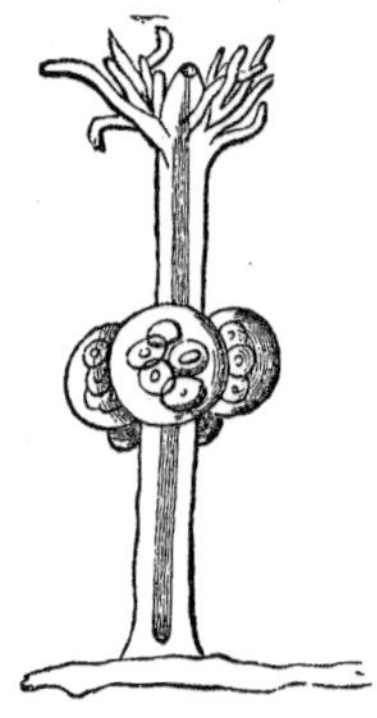

FIG. 15.

becoming a new generation; the sexual organ has become the sexual animal.

Now as the individual development of the Cladonema, and other Medusæ similarly propagated, corresponds with the systematic series of the Medusa polypes, the only reasonable and credible explanation of the ontogenesis of those Medusæ in which heterogenesis occurs, is that, in them, the historical development of the genus has become fixed. Neither the egg nor the hen were created. Before the delicately tinted Medusæ populated the primæval ocean in lonely splendour, the Medusa polypes on the constantly changing shores were the sole representatives of the still infant class. Why single genera, like the Hydractinia, remained strictly conservative while others in various degrees paid homage to progress, whether and how the struggle for existence and survival of the fittest were here concerned, it is certainly impossible to prove in the individual species. But the general impression is decisive, and also the circumstance that the theory is consistent with the facts.

The evolutionary history of the intestinal worms leads to the same reflections and results. These animals, widely differing in their structure, were either created in or with their hosts, or else they have become habituated to them in a natural and direct manner. We may surely disregard the third alternative, that they were led by an innate "obscure impulse." According to our doctrine, the worms now passing the whole or a portion of their lives as parasites on or in other organisms, are descended from free and independent animals, and the periods occurring in their development, during which parasitic life is exchanged for independent phases, signifies a reversion taking place systematically in all individuals to the once permanent condition of their progenitors. Of the Trematoda or Flukes, and Cestoda or Tapeworms, belonging to the class of the Platelmintha Suctoria, the latter have diverged the most from their starting-point; their adaptation to life within other animals has rendered the alimentary canal superfluous, and their generations and transformations hence point less to their progenitors than is the case with a number of other Trematoda, with which many anatomical characters prove them to be closely related. Both, moreover, share the characters of their class with the free-living Turbellaria. From such as these, that is to say, from forms approximate to the present Turbellaria, the Trematoda and Cestoda must be descended, and with this agrees the free roving phase which the larva of the Fluke (Distomum) undergoes as the so-called Cercaria, and previously as a rotating spherical body.

Many of the ciliated Nematoids, or thread-worms, too,—the division which includes the Ascarides among others,—have in their infancy a stage of independent life, during which they cannot be distinguished from the infantine forms of their more numerous kindred, which never adopt a parasitic life, and chiefly inhabit the sea. The transition to parasitism, as recapitulated by ontogenesis, was nothing more than an extension to a new territory offering advantages of nutriment; and on this point it is highly instructive to compare the Nematodes with the systematic series of the leech-like Suctoria (Trematoda), so excellently described by Van Beneden. We here find all the transitions from independent predatory genera to others occasionally parasitic, and again from these to others which on leaving the egg immediately attach themselves for life. Here, as elsewhere, parasitism seems an adaptation to new habitats, which is recorded in the biography of the individual with a reminiscence of the previous form.

The circumstances of the parasitic worms are repeated by the parasitic Crustacea, as, moreover, a probably primordial form of the crab family is preserved in the metamorphoses of several orders of this large and diversified, though coherent class. The larva, which, it may safely be assumed, approximates closely to the primordial form, was at one time taken for an independent genus and received the name of Nauplius. Hence a Nauplius phase is spoken of, which obtains especially among the lower Crustacea, the Copepoda, parasitical Crustacea and Cirripedes, and the remarkable Rhizopoda connected with them; but is not wanting in the highest order, the decapodous stalk-eyed crab. We shall later have to make acquaintance with the so-called curtailed development which among the crabs has been adopted by the decapods, and it was formerly supposed by all. Were this actually the case, we should still, by analogy, infer their connection with the other orders repeating the Nauplius phase in the course of their development; but it was a welcome

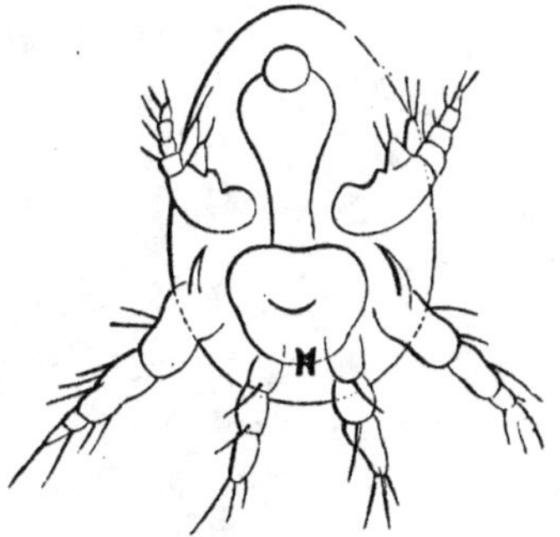

FIG. 16.

discovery of Fritz Müller's that a shrimp (Peneus) still begins its development as a Nauplius; whereas all the other members of the order, as far as they are known, leave the egg in the higher Zoea phase (p. 50). As of the hundreds of stalk-eyed crabs, scarcely a dozen have been hitherto examined as to their development, it will not be doubted that, with regard to the Nauplius phase, some resemble the Peneus of the Brazilian coast. But even were this case to prove unique in the order, it would suffice as a living witness of the connection

FIG. 17. Axolotl.

between the presence of the decapods and the primordial crabs. There can be no other view of this subject. The Nauplius phase in the development of the Peneus is either a shining testimony in favour of the doctrine of Descent, or a senseless paradox.

After what has gone before, the transformation of the Amphibians needs no elucidation. Their predecessors were water-breathers, whose form and mode of life are more faithfully preserved by the long-tailed Amphibians, the tritons, and salamanders, than by the frogs. In our tritons, sexual maturity not rarely commences in the larval state, hence in a phase which was definitive in the progenitors of the present genera. There is, indeed, one species, the Mexican Axolotl, which normally propagates itself during the larval phase. Auguste

FIG. 18. Amblystoma.

Dumeril's observation is highly interesting, that of the thousands of Axolotls that he bred at Paris, some few advanced beyond the grade of development hitherto known in them, *i.e.* they lost their gills, changed the shape of their bodies not inconsiderably, and from gill-breathers, and aquatic animals became lung-breathers and terrestrial animals. It needs further observation to ascertain whether (what is, however, very improbable), in their home, all Axolotls, after having propagated themselves in their larval state, undergo the metamorphosis into salamander-like animals (Amblystoma), or whether the transfer to Europe and the consequent entire change of the circumstances of life gave the impulse to a progressive transformation of these few individuals, which, by the continuance of these conditions, would in future generations extend to more and more individuals, and finally become the characteristic of a new species.

The examples of Ontogenesis, or individual development, hitherto examined, had the peculiarity that the sexual animal does not issue directly from its egg like the Phœnix from its ashes, but had to pass through various forms and existences in which the progenitors of the species again become alive and palpable. We must now inquire how this development is related to that form of reproduction which the systematizers, completely in accordance with the facts, yet without any corresponding meaning, have termed "direct development," or "development without heterogenesis or metamorphosis?" The ciliated embryos of many Medusæ are not converted into polype-like intermediate forms, but pass directly into Medusæ. The greater number of higher crabs do not leave the egg as Nauplia, but as more or less perfect decapods. The bird, the mammal, and man are all at birth "similar to their parents." Considering that the processes of heterogenesis are in themselves by no means advantageous to or "in harmony with design"—we have only to remember the fate of the tapeworm's eggs—that by the larval state the period of infancy and weakness is prolonged, and the period of maturity and efficient care for the continuance of the species delayed, it follows that curtailments and reductions, consequent on adaptation have, as advantageous modifications, a prospect of perpetuation. As in Amphibians the prolongation of the larval phase may be effected by natural circumstances and artificial experiments, so in like manner a compression of the phases of transformation, and a general curtailment of the metamorphosis is imaginable. In the class of Amphibians we have, in fact, several examples of curtailed and modified metamorphosis which bridge over the apparent chasm between development with and without transformation, and render direct development comprehensible as being gradually acquired. Amphibians will endeavour to extend themselves wherever they are invited by a sufficient supply of insects, and the black salamander of the mountains (Salamandra atra) has even overcome the impediment which might have been deemed insurmountable, the absence of water for its larvæ. It does not lay its eggs like its congeners, but only two are received into the oviduct, and the fluids secreted from its walls replace the marsh to them and to the larvæ which emerge from them. Here, and not when separated from the parent, do the gills make their appearance, while the other eggs, gradually following, are devoured by the hungry larvæ. The metamorphosis of the black salamander, respecting which, unluckily, no recent investigations have been made, thus takes place within the parental body, and there is no difficulty in imagining the acquisition of this peculiarity by the necessity of adaptation. If the mode of life of the marsupial frog, which carries its young in a membranous fold of the back, and the Surinam toad, of which the larvæ live singly in the chambers of a kind of honeycomb on the back, were better known than they are, we should assuredly arrive at the same results as with the black salamander. In the absence of other knowledge, the observations of M. Bavey, Marine Pharmaceutist at Guadaloupe, first published in 1873, are of the highest importance.[66] A frog of those parts (Hylodon Martinicensis) goes through its whole metamorphosis in the egg. In the egg it has gills and tail; and from the brief remark that the island contains only rapid running streams, and nowhere stagnant waters or marshes, it appears that this is also a case in which adaptation modifies and curtails development.

If, after this introduction, we now examine the so-called direct development with more attention, it may in every way be compared to the metamorphosis of the Hylodes of Guadaloupe. Direct development is a transformation in the ovum; and in the cases in which it occurs, the phases of embryonic development are repetitions, more or less distinct, of the historic development of the family. We will only particularize in the embryonic life of the Vertebrata (in which metamorphosis does not take place), some phases that are stages of curtailed transformation, and recapitulate the permanent condition of their progenitors. It has been repeatedly mentioned that in all vertebrate animals, the vertebral column is first laid out as an unsegmented cord and an unsegmented sheath for the spinal cord. This is the permanent state of the lower fishes. In the higher Vertebrata also, the brain at first consists of vesicles, lying one behind the other, which is the persistent form of the lower groups. The embryonic heart of mammals and birds begins in the form of a tube, and subsequently acquires the communications between the chambers, which in the reptiles never close. In the Amphibians, the branchial arches really bear gills during the larval state. They are not wanting in the embryos of reptiles, birds, and mammals, any more than the fissures through which, in fish and the larvæ of Amphibians, the water passes off after being inhaled. Must we again set forth the only possible explanation of these facts?

Before referring to the phenomena which testify the emanation of families from a common root, we will cite one of the most important evidences of recent times, which traces the genesis of species through a great geological period, and exhibits in detail the relations of the development of the individuals to that of the species, genus, and family. We mean L. Würtenberger's contribution to the geological evidence of the Darwinian theory, to which we have already appealed (p. 97). It relates to the two families of Ammonites, the Planulata and Armata; of which, according to Würtenberger's researches, the latter are developed from the former, as the ribs of the Planulata gradually pass into the spines of the Armata. Of special interest to us are the following passages of the preliminary communication on the discoveries obtained from thousands of specimens, and which will probably not be made public, with all the vouchers, for some years to come. "It gave me particular pleasure," says Würtenberger, "when, after divers careful comparative studies, I at last detected an interesting and simple conformity to law in the variations of the Ammonites. Namely, on the first appearance of a modification which subsequently attains essential importance in an entire group, it is only slightly indicated on a portion of the last convolution. Towards more recent deposits, this modification is more and more plainly shown, and then advances, following the spiral course of

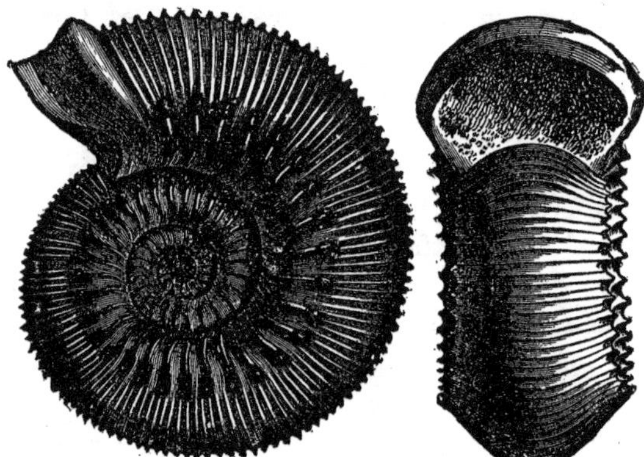

FIG. 19. Ammonites Humphresianus. A form analogous to the Planulata.

the shell; that is to say, it gradually takes possession of the central turns also, as we trace the forms to higher strata. This reproduction in younger stages of life of modifications first occurring at a more advanced age, makes but slow progress, so that we see the older forms repeated with great persistency in the central turns. Frequently a modification of this sort has taken possession of only a small part of the convolutions, when a new one already appears at the outside, and follows the first. Thus searching through the strata from below upwards, we see modification after modification beginning at the outer part of the Ammonites, and advancing towards the centre of the discs. The innermost convolutions often resist these innovations with great persistency, so that we usually find upon their surface several of these states of development closely compressed, as the shell of the individual Ammonite begins with the old morphological type, and then adopts the modifications in the same order in which they follow in vast periods in the geological development of the groups concerned."

"The Ammonites," he says moreover, "thus obtain at an advanced and maturer age—only when they have gone through the development inherited from their parents, and as much as possible in the same manner as their parents—the power of modifying themselves in a new direction, that is to say, of adapting themselves to new conditions; yet these modifications may then be transmitted to the offspring, so as to appear in each subsequent generation a trifle earlier, until this phase of development in its turn characterizes the greater portion of the period of growth. But this last and longest phase of development scarcely ever suffers itself to be supplanted by new ones, formed in like manner; heredity operates so powerfully, that a period of development thus once predominant, is repeated in the infancy of the Ammonites, even though but slightly indicated. Hence in an individual Ammonite from a recent stratum, the periods of development compressed and forced back upon the innermost convolutions, must appear in the same succession in which they wrested the dominion from one another. It is extremely interesting to study the development of the Inflata of the upper white Jura, which follow the Ammonites liparus (whose externally visible convolutions display only one row of spines), and carefully break off convolution by convolution. Towards the middle there is a region in which there are always two rows of spines ; nearer the centre the innermost row disappears; soon afterwards the outer one also ; and the nucleus, some millimetres in diameter, now appears for about half a turn as a Planulatum, with distinct ribs, which, towards the beginning, likewise disappear. Thus even the Planulate ribs, which prevailed among the Liassic ancestors of these Inflata, and were supplanted by the spines as early as in the brown Jura, still distinguish these later and essentially modified descendants during a short period of their youth."

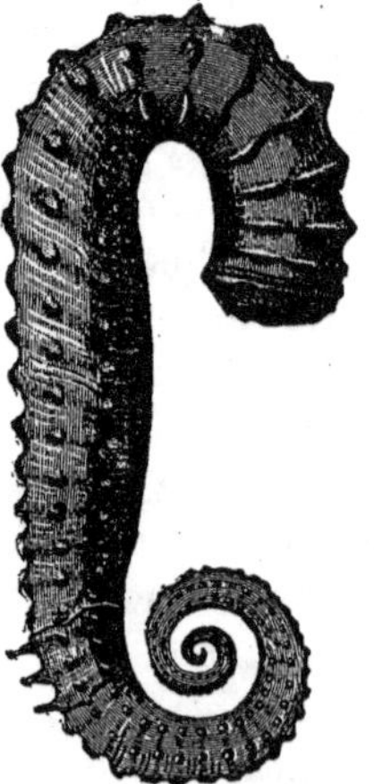

FIG. 20. Ancyloceras.

Würtenberger further shows how these relations can be simply explained by the Darwinian theory alone; "without it we should have only an extraordinary problem."

It was natural to test the applicability of the theory of selection also on the forms allied to the Ammonites, such as the Ancyloceras; namely, the genera in which the convolutions do not touch and partially conceal one another, as in genuine Ammonites, and which, as late comers and side shoots of the group, seemed destined to decay. Selection and decay? Würtenberger shows how the abandonment of contact in the convolutions was to the spinous Ammonites an advantage which would be established by selection. If other palæontologists consider the fluctuations of form accompanying the relaxation of the closed spiral as evincing the decline of the group, no contradiction seems to be implied, for what was originally used as an advantage by natural selection, proved injurious in its consequences.

As we have seen, the earliest states are obliterated to

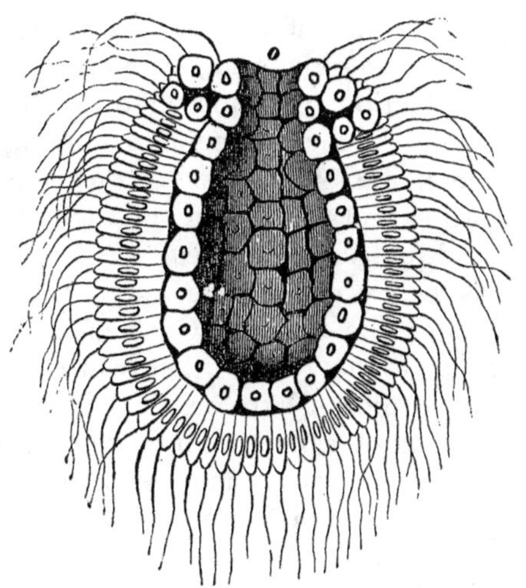

FIG. 21.

such a degree by curtailment of development that the indication of the nature of the progenitors continually diminishes. But our theory necessarily leads to the conviction that the families within which we have as yet been able to compare Ontogenesis with Philogenesis, constantly approximate in their origin, and vindicate the expectation that at least here and there, in the individual development of single representatives of the various families, witnesses of their common derivation should come to light. This likewise occurs, and to such a degree that in the earliest larval stages a link is established between the lowest and the highest animals. If a number of groups of the lowest living beings, in which the various vital functions of nutrition, irritability, motion, and reproduction are supplied by amorphous protoplasm,—if these be separated, as by Haeckel, into a neutral kingdom, owing to the absence of sexual reproduction, we must likewise agree with him in attributing to the Spongiadæ ranking next to the Protista, the name of animals, on account of their sexual propagation and the nature of their embryonic development and first larval phases.

Haeckel has bestowed on one larval stage of the calcareous sponges the title of Gastrula, wherein the animal represents a sac, or, in other words, a stomach provided with a mouth-like orifice. The walls are formed of two rows of cells, the outer one consisting of ciliated cells ; that is to say, each cell is furnished with a long filament. At the orifice of the sac, the outer row merges into the inner one, and from these two membranes the body of the sponge is constructed in a definite manner. Now, if this Gastrula larva reappears in the Cœlenterata, Polypes, and Medusæ, in which the gradual development from the two membranes, the entoderm and ectoderm, into the most complex forms has long been known ; and if, as Haeckel has further shown, the osculum, or larger opening of the spongiadæ may be closely compared with the mouth of the polype and medusa, and the great central cavity of the sponge with the stomach of the others, of the canal system with the canals and cavities of the Cœlenterata,—then, in combination with the host of other facts, implying and supporting the doctrine of Descent, the inference is inevitable that in the Gastrula we have a testimony of the consanguinity of the Spongiadæ and Cœlenterata. But this Gastrula reappears in the Holothuria; hence in the Echinoderms, in the Sagitta, in the Ascidians, which will be more narrowly examined in the pedigree of the Vertebrata, and finally in the Lancelet; and we, therefore, hold ourselves justified in regarding this coincidence of the earliest states of development in different families, as the remnant of the common root, which in other families, as in the Articulata, for example, has been lost in the curtailment of development. The significance of the "germinal membranes" in the Vertebrata was recognized even by Pander, and in the suggestive works of v. Baer; the extension and application of this observation to the whole animal kingdom, for which we are especially indebted to Kowalewsky, marks one of the greatest advances in the science of comparative development.

The reader unacquainted with the detailed researches of our science, has already been called upon to observe that there are opponents of the theory of selection, such as Owen, who nevertheless accept the doctrine of Descent as incontestable. Even rejecting natural selection, the parallelism of Ontogenesis with Phylogenesis may also be brought into the natural connection maintained by us, on the assumption of an unnatural or supernatural guidance which converts this apparently natural unity into a miracle. Quite recently, A. Braun has pointed out the accordance of the botanical system, and therewith of palæontological succession, with the development of the individual plant, when he says :[67]—"In the further elaboration of the natural system, the gradation of the vegetal kingdom, and, at the same time, the relation of the system to the history of development, becomes more and more spontaneously and incontrovertibly manifest. The Acotyledons are verified as Cryptogams, as they were already considered by the old botanists of pre-Linnæan times, and their relation to the Phænogams is thus more clearly pronounced. The Cryptogams are separated into two essentially different divisions, in which gradation is likewise distinctly pronounced (cellular and vascular Cryptogams, Thallophytes and Kormophytes); between the perfect Phænogams and the Cryptogams an intermediate grade has been shown, that of the Gymnosperms. But most important of all is the circumstance that the four chief grades ascertained in the vegetal kingdom accurately correspond with the grades of development occurring in the individuals of all the higher plants;—the germ, the vegetative stem, the blossom and the fruit." But why this parallelism is to be most important of all, if it is not to lead us to the knowledge of true causality, is beyond our comprehension. We can well imagine that the "inherent causes" and the "Principle of Perfection" may be welcomed as the *refugium ignorantiæ*, but not that they can really satisfy inquiry. For our own standpoint, the accordance of the results of botanical investigation must be extremely important, but it is for the palpable reason that the theory thereby gains the support and corroboration of another great series of facts.

If the accordance of the evolution of families has once been followed up to the Gastrula, we shall not pause there, but must regard the similarity of the sperm corpuscules and germ cells from the Spongiadæ to the Vertebrata as a primordial common property, connecting the animal and vegetal world; and prior to the acquisition of which, only those modes of reproduction took place which have been maintained among Protista and in heterogenesis.

As the common basis of sexual reproduction in the various families argues a common origin, asexual reproduction, directly connected as we have seen it to be with sexual propagation, by means of unfecundated eggs and germs, leads us constantly further towards the beginning of life. But the cell furnished with a nucleus and sheath is inseparable from the protoplasmic corpuscule devoid of nucleus or sheath, on the growth and fission of which rests the reproduction of the lowest living beings.

Their origin from inorganic matter, as we have set forth above, is a postulate of sound human understanding. To this beginning we are led, not, as the opponents of the doctrine of Descent are wont to say, by a dogmatic after-philosophy, but by the unprejudiced consideration and computation of the facts of individual development.[68]

X.

The Geographical Distribution of Animals in the light of the Doctrine of Derivation.

ALTHOUGH ever since the century of the great geographical discoveries, material has been accumulating for a geography of plants and animals, the foundations of scientific botanical geography (apart from George Forster's observations) were first contained in Humboldt's celebrated "Ideas on the Physiognomy of Plants" (Ideen zu einer Physiognomik der Gewächse). It is the first description of vegetal forms, comprising the entire area of the earth, and the manner in which, singly or combined, they lend a characteristic impress to the landscape of their region of distribution, and again on their side harmonize with the other factors of the scene. The celebrated founder of Climatology, who circled the terrestrial globe with lines of equal temperature, of equal inclination and declination of the magnetic needle, and divided it into dry and rainy zones, knew better than any of his contemporaries that the animal and vegetal world depended on all these factors. Yet neither he nor his followers, before Darwin, rose higher than the description of Nature, which had already checked Buffon in his grand picture of Nature, "Les Epoques de la Nature."

A natural result of the extraordinary extension of

the geographical horizon and the profundity of special research was the more careful ascertainment of the regions of distribution of animal and vegetal families, and of their more prominent species, in which, as we have already said, either no questions were asked as to the causes of distribution, or the matter was facilitated, as by Louis Agassiz, who did not, like Linnæus, derive each species from a pair, but supposed them to be created in suitable numbers of individuals in their own regions of distribution. It cannot be expected that any solution was hereby given to the questions which now force themselves upon us, such as why, under like natural conditions, like species are not always to be found, and conversely? Why very similar species frequently appear under external conditions entirely dissimilar? What is to be thought of the mutual relations of the so-called vicarious forms? &c.

As Rütimeyer has recently observed, in his excellent treatise "On the Derivation of the Animal World of Switzerland" ("Ueber die Herkunft der schweizerischen Thierwelt"[69]), Buffon had already remarked the repetition of the African in the American fauna; how, for example, the lama is a juvenescent and feeble copy of the camel; and how the puma of the New represents the lion of the Old World. Still, by the mere word "representative" or "vicarious form" nothing is gained, and a true apprehension of these facts is obtained singly and solely if we meet the inquiry with the assumption that camel and lama, puma and lion, are of common derivation, and that their diverse development was in the lapse of time favoured and determined by the separation of the habitats of their progenitors.

Another example of so-called vicarious or "analogous" species, affording an easier basis for induction, is provided by the comparison of the snails of Southern Europe, and especially of Spain, with those of North Africa, on which we are indebted to Bourguignat for some excellent observations. In accordance with other botanical and zoological facts, he has established that the shell fauna of Spain and North Africa forms a whole, so that the Algierian snails appear a mere appendage to those of Southern Europe, nothwithstanding the separation by the Straits of Gibraltar. Now it is proved that, in geologically recent times, this region of North Africa was in fact a peninsula of Spain, and that its union with Africa was effected on the north by the rupture of the Straits of Gibraltar, and on the south by an upheaval to which the Sahara owes its existence. The shores of the former Sea of Sahara are still marked by the shells of the same snails that live on the shores of the Mediterranean. But all North African species are not identical with those of Spain; of many African sorts, only "analogous" species are found on our side. Now if certain Spanish species do not themselves occur in Africa, but are yet replaced by very similar forms, our standpoint at once connects with the otherwise unmeaning word "analogous" species the idea of the common derivation of the forms replacing one another, and of the local variations superinduced by isolation and altered conditions.

A severe test is applied to those who believe that species were separately created, by the air-breathing land snails (pulmo-gasteropoda), when it is seen that in isolated islands and island groups these earth-bound animals, migrating with so much difficulty, have attained an extraordinary diversity. In the Madeira Islands, 134 species of pulmo-gasteropoda were reckoned about ten years ago, of which only 21 were to be found in the Africo-European fauna. These and the 113 other species are mostly confined to narrow districts and single valleys. Are we to suppose that the 113 species for Madeira, and the 21 species for Madeira and Africa with Europe, were each separately created? Must we not much rather infer that a connection at one time existed between Europe and the present island group of Madeira, and that these 21 species remained what they were before the separation; while from unknown species still appearing in analogous forms upon the continent emanated the remarkable profusion of new species? They, and their comrades on other isolated islands, were spared a conflict many sided, and they doubtless afford a favourable example of Wagner's law of migration, as with the difficulties of locomotion, and the improbability of a large subsequent arrival, the secluded individuals, under even slightly different influences, had had a prospect of diverging from the parent species.

The unscientific opinion, that under like, or nearly like, external conditions, like or similar organisms were created in great numbers, receives a severe blow by the perception that the direct reverse has frequently occurred. Why has America no horses in the present era, although it is proved that the horses introduced, thrive capitally? It is not necessary for us to explain why the fossil horses which existed in America, as well as in the Eastern hemisphere, became extinct without leaving any progeny—we do not know the cause, though we may yet be able to fathom it; but in this and all similar cases the adherents of the doctrine of Creation must confess the inadequacy of their theory of belief.

Our exposition has shown that the species now extant are the progeny of organisms previously existing; the present apportionment on the earth is therefore a consequence of the distribution of the progenitors of the present organisms, and of the manifold displacements of land and water by which they were indirectly or directly affected. We cannot hope ever to picture to ourselves a faithful representation of the perpetual transformations of the surface of the earth. Only, if this could be accomplished, and if we, moreover, had an accurate register of the animals at each period inhabiting the former islands, continents, and oceans—only then could the distribution of the present organisms be thoroughly fathomed and established. But in thus acknowledging the incompleteness of our statistical means, we are at least able to lay down with certainty the course of inquiry. We must, in the first place, proceed in the method of the older vegetal and animal geography to ascertain the natural limits and regions of distribution; and, secondly, to collate these facts with the facts of the distribution of the former progenitors of the present animate world as it was determined by the geological conditions of those times. It is needless to say that Darwin has furnished the outlines for this work also. But among his followers two are specially worthy of distinction: Wallace, with his researches on the Malay Archipelago,[70] abounding in subtle observation;

and Rütimeyer, in his treatise already cited. In what follows we may essentially adhere to the latter.

Our knowledge of the regions of distribution of the animal world is still extraordinarily deficient. What do we know, for instance, of the occurrence of marine animals? Few years only have elapsed since the depths of the sea were rendered accessible to research, and the result has almost entirely upset our earlier notions of the geological significance of the sea-bottom and its habitability. After the strong impulse given by Maury to the investigation of the physical condition of the sea, we are now occupied in ascertaining the submarine temperatures and currents, the constitution of the sea-bottom, the occurrence of deep-sea organisms, and the conditions of their existence. We are therefore just beginning to collect the material for a future geography of marine organisms. Among terrestrial animals, certain groups of which the actual distribution can be defined, are useless for our general purpose.

Butterflies, for instance, which are an easy prey to currents of air, defy geological barriers, and, above all, that important partition which from the tertiary era has been erected, or rather excavated in the bottom of the sea, between Australia and India.[71] It is the same with bats, and also with migratory, predatory, and aquatic birds; while, as Wallace shows, the other orders of this class are in tropical regions very reliable and stable inhabitants of their often limited districts, seemingly suggestive of migration. Exclusive of these, there remains therefore little more than the Mammalia, whose extraction may be inferred with certainty from a comparison of their present cantonments (Cantonirung),

—an expression which we borrow from Rütimeyer,—with the encampments of their former kindred, whence are derived general points of view as to the causes of the present geographical apportionment of organisms.

If in the preliminary establishment of facts we therefore confine ourselves to the Mammalia, exclusive of whales and bats, a superficial survey is enough to show that not only single species, but families also, have each a certain region of greatest density of occurrence, a focus of distribution, and that from thence radiations have taken place according to the convenience and fitness of the territory. Lion and tiger, elephant and camel, range over a definite area; the monkeys of the New World differ from those of the Old World not only geographically, but also in family characteristics. Marsupials are chiefly concentrated in Australia; sloths and armadilloes in South America. And these examples, easy to multiply, indicate how individuals of widely dispersed species, and the species themselves, emanated from single points of the earth's surface and flowed over the territory of distribution now occupied. When to this observation is added the other, that in past eras also the same groups had the same centres of distribution,—for instance, Brazil not only harbours sloths and armadilloes now, but was once peopled by more numerous and partly colossal species of these families, and Australia has furnished the most numerous and important fossil remains of Marsupials,—the cognizance of this persistent localization becomes very significant, and we account for the "repetition" of these forms by derivation.

Now if the centres of distribution, at the first glance

extremely numerous, can be brought into closer union and reduced to the smallest number possible, as by our theory the Mammalia have but one point of derivation, and if we can herewith harmonize the geological succession of the organisms examined, or, in other words, harmonize the horizontal distribution with the vertical or historical sequence, animal geography will then approach the solution of its task. Wallace and Rütimeyer's works are therefore an important advance, as the former has given detailed evidence that the fauna of the complex and extensive Australio-Indian Archipelago is by no means self-dependent, but consists merely of offshoots of the continents; and the latter, in a grand survey of the entire surface of the earth, has reduced the centres of distribution to the simplest proportions as yet possible.

The comparison of insular and continental faunas is naturally of great interest. For should it appear that, with respect to the animal world, islands are one and all mere appendages of the continents, the problem would at once be vastly simplified. If we follow Peschel's luminous exposition of the origin of islands,[72] we have first to deal with the fragments of continents. A great number of islands, such as Great Britain and the great Asiatic islands, may be recognized at once as fragments of still existing continents. On the other hand, Madagascar and the Seychelles are not, as might be conjectured, a segment of Africa, but the remnant of a former continent very peculiar in its flora and fauna. Other islands originate either from submarine volcanoes or from corals, and in the latter case the structure is founded on sinking land. It naturally follows that on volcanic and coral islands only such animals will be encountered as reached them by swimming or flying. The presence of Mammals pre-supposes human agency or extraordinary accidents. The older the islands, the richer are they in organisms. Islands detached from continents will, on the contrary, be rich in proportion as they are recent, of which Great Britain bears witness. The more divergent is their fauna, the longer must be the time which has elapsed since their separation. Thus, for instance, we may view the relations of Tasmania and Australia; and if New Zealand was ever connected with the old Australian continent, the separation occurred at an epoch so remote that it throws no light upon the physiognomy of the animal world of New Zealand, and *vice versâ*.

In the account of his travels in the Malay Archipelago, Wallace has given a pattern of animal-geographical research. Years before, G. Windsor Earl had pointed out that the great islands of Sumatra, Borneo, and Java, are connected with the Asiatic continent by a shallower sea; while a similar shallow sea assigns New Guinea and several adjacent islands to Australia, with which they have a common characteristic in the Marsupials. Wallace has defined this partition more minutely with a line marked by a deeper submergence of the sea-bottom. It is drawn below the Philippine Islands, and, having Celebes to the south, passes through the straits of Macassar and separates the two small islands of Bali and Lombok. We will now follow Wallace's description ("Malay Archipelago"), with various omissions.

"It is now generally admitted that the present distribution of living things on the surface of the earth is mainly the result of the last series of changes that it has undergone. Geology teaches us that the surface of the land and the distribution of land and water is everywhere slowly changing. It further teaches us that the forms of life which inhabit that surface have, during every period of which we possess any record, been also slowly changing. As to the Malay Archipelago, we find that all the wide expanse of sea which divides Java, Sumatra, and Borneo from each other, and from Malacca and Siam, is so shallow that ships can anchor in any part of it, since it rarely exceeds forty fathoms in depth: and if we go as far as the line of a hundred fathoms, we shall include the Philippine Islands and Bali, east of Java. If, therefore, these islands have been separated from each other and the continent, by subsidence of the intervening tracts of land, we should conclude that the separation has been comparatively recent, since the depth to which the land has subsided is so small.—But it is when we examine the zoology of these countries that we find what we most require—evidence of a very striking character that these great islands must have once formed a part of the continent, and could only have been separated at a very recent geological epoch. The elephant and tapir of Sumatra and Borneo, the rhinoceros of Sumatra and the allied species of Java, the wild cattle of Borneo and the kind long supposed to be peculiar to Java, are now all known to inhabit some part or other of Southern Asia. None of these large animals could possibly have passed over the arms of the sea which now separate these countries, and their presence plainly indicates that a land communication must have existed since the origin of the species. Among the smaller mammals, a considerable portion are common to each island and the continent; but the vast physical changes that must have occurred during the breaking up and subsidence of such extensive regions have led to the extinction of some in one or more of the islands, and in some cases there seems also to have been time for a change of species to have taken place. Birds and insects illustrate the same view, for every family, and almost every genus of these groups found in any of the islands, occurs also on the Asiatic continent, and in a great number of cases the species are exactly identical. Birds offer us one of the best means of determining the law of distribution; for though at first sight it would appear that the watery boundaries which keep out the land quadrupeds could be easily passed over by birds, yet practically it is not so; for if we leave out the aquatic tribes which are pre-eminently wanderers, it is found that the others (and especially the Passeres, or true perching birds, which form the vast majority) are generally as strictly limited by straits and arms of the sea as are quadrupeds themselves. As an instance, among the islands of which I am now speaking, it is a remarkable fact that Java posesses numerous birds which never pass over to Sumatra, though they are separated by a strait only fifteen miles wide, and with islands in mid-channel. Java, in fact, possesses more birds and insects peculiar to itself than either Sumatra or Borneo, and this would indicate that it was earliest separated from the continent; next in organic individuality is Borneo; while Sumatra is so nearly identical in all its animal forms with the peninsula of Malacca, that we may safely conclude it to have been the most recently dismembered island.

"The Philippine Islands agree in many respects with Asia and the other islands, but present some anomalies to indicate that they were separated at an earlier period, and have since been subject to many revolutions in their physical geography.

"Turning our attention now to the remaining portion of the Archipelago, we shall find that all the islands, from Celebes to Lombock eastward, exhibit almost as close a resemblance to Australia and New Guinea as the Western Islands do to Asia. It is well known that the natural productions of Australia differ from those of Asia more than those of any of the four ancient quarters of the world differ from each other. Australia, in fact, stands alone; it possesses no apes or monkeys, no cats or tigers, wolves, bears, or hyenas, no deer or antelopes, sheep or oxen, no elephant, horse, squirrel or rabbit; none, in short, of those familiar types of quadruped which are met with in every other part of the world. Instead of these, it has Marsupials only, kangaroos and opossums, wombats and the duck-billed platypus. In birds it is almost as peculiar. It has no woodpeckers and no pheasants, families which exist in every other part of the world; but instead of them it has the mound-making brush-turkeys, the honeysuckers, the cockatoos, and the brush-tongued lories, which are found nowhere else upon the globe. All these striking peculiarities are found also in those islands which form the Austro-Malayan division of the Archipelago.

"The great contrast between the two divisions of the Archipelago is nowhere so abruptly exhibited as on passing from the island of Bali to that of Lombock, where the two regions are in closest proximity. In Bali we have barbets, fruit thrushes, and woodpeckers; on passing over to Lombock these are seen no more, but we have abundance of cockatoos, honeysuckers, and brush-turkeys, which are equally unknown in Bali or in any island further west. The strait is here fifteen miles wide, so that we may pass in two hours from one great division of the earth to another, differing as essentially in their animal life as Europe does from America.* If we travel from Java or Borneo to Celebes or the Moluccas, the difference is still more striking. In the first, the forests abound in monkeys of many kinds, wild cats, deer, civets and others, and numerous varieties of squirrels are constantly met with. In the latter, none of these occur, but the prehensile-tailed cuscus is almost the only terrestrial mammal seen, except wild pigs, which are found in all the islands, and deer (which have probably been recently introduced) in the Celebes and the Moluccas. The birds which are most abundant in the Western islands are woodpeckers, barbets, trogons, fruit-thrushes, and leaf-thrushes; they are seen daily, and form the great ornithological features of the country. In the Eastern islands these are absolutely unknown, honeysuckers and small lories being the most common birds; so that the naturalist feels himself in a new world, and can hardly realize that he has passed from the one region to the other in a few days, without ever being out of sight of land.

"The inference that we must draw from these facts is undoubtedly that the whole of the islands eastwards,

* This is too vaguely expressed. It would be nearer the mark to say, as Europe does from South America. (O. SCHMIDT.)

beyond Java and Borneo, do essentially form a part of a former Australian or Pacific continent, although some of them may never have been actually joined to it. This continent must have been broken up not only before the Western islands were separated from Asia, but probably before the extreme south-eastern portion of Asia was raised above the waters of the ocean; for a great part of the land of Borneo and Java is known to be geologically of quite recent formation; while the very great difference of species, and in many cases of genera also, between the productions of the Eastern Malay islands and Australia, as well as the great depth of the sea now separating them, all point to a comparatively long period of isolation."

"It is interesting to observe among the islands themselves how a shallow sea always intimates a recent land connection. The Aru islands, Maisol and Waigiou, as well as Jobic, agree with New Guinea in their species of mammalia and birds much more closely than they do with the Moluccas, and we find that they are all united to New Guinea by a shallow sea. In fact, the 100-fathom line round New Guinea marks out accurately the range of the true Paradise birds.

"It is further to be noted—and this is a very interesting point in connection with theories of the dependence of special forms of life on external conditions—that this division of the Archipelago into two regions characterized by a striking diversity in their natural productions, does not in any way correspond to the main physical or climatal divisions of the surface." We will further quote only the following: "Borneo and New Guinea, as alike physically as two distinct countries can be, are zoologically wide as the poles asunder ; while Australia, with its dry winds, its open plains, its stony deserts, and its temperate climate, yet produces birds and quadrupeds which are closely related to those inhabiting the hot, damp, luxuriant forests which everywhere clothe the plains and mountains of New Guinea."

Wallace gives the most specific proofs that, as the parts of this Archipelago approach one another like separated extremities of two continents, they bring with them two entirely different fauna. Similarly, the Mediterranean and West Indian Archipelagos are devoid of any peculiar character, and are completely dependent on the adjacent continents for their animal life and vegetation. We have already discussed Madeira and its land snails. Insular faunas therefore do not require the hypothesis of more centres of creation than are offered by the continents; and Rütimeyer has endeavoured to trace the extraction of birds and mammals to two centres of derivation. A great series of animal-geographical facts is explicable only on the hypothesis of the former existence of a southern continent, of which the Australian mainland is a remnant. The present Marsupials are concentrated in Australia. Their occurrence in the south-western portion of the Malay Archipelago, including New Guinea, seems like a radiation from that centre. No single token makes it appear that the Marsupials existing in former periods in the northern hemisphere, from the Jura forwards, had migrated to meet those which were pressing on from the southern continent towards the equator. Only as to the opossum, so widely extended in South America, could a question arise, which is however solved by the examination of a host of congeners, one and all alien to the population predominant in America, and indicating importation probably in the Tertiary period; unless it be assumed, with Rütimeyer, "that implacental mammals were created out of Australia as well as in it."

Among the first to be mentioned are the wingless birds, that is, those which are anatomically and systematically connected, and which we now find scattered over continents and some of the larger islands. The cassowary of New Holland and America, the extinct giant birds of Madagascar and New Zealand, the African ostrich, which has advanced from the south northwards, cannot have originated in their present isolation. The same considerations are forced upon us by the mammals named Bruta by Linnæus, and by modern zoologists termed Edentata, by reason of their imperfect dentition, among which, accepting the latter definition, must be included the Ornithorhyncus, or duck-mole of Tasmania. These duck-moles incontestibly occupy the lowest grade among the mammals now extant; but the other true Edentata are no less alien to the higher orders, and their occurrence in South America on the one hand, and in South Africa and South Asia on the other, as well as the impossibility of tracing them from a common centre in the northern hemisphere, points to the vanished land of the south, where perhaps the home of the progenitors of the Maki of Madagascar may also be looked for.

"Or," says Rütimeyer, "does the hypothesis of a Polar land, once possessing an abundance of animal life, partly covered by the ocean and partly by a coat of ice, appear an unfounded assumption to us who now witness the elevation of a similar frozen surface in the northern hemisphere, and are surrounded in the Alps by a still existing—in our glacial drift by a scarcely vanished—arctic scene? Or need the conjecture that the almost exclusively graminivorous and insectivorous Marsupials, sloths, armadilloes, ant-eaters, and ostriches, once possessed an actual point of union in a southern continent, of which the present flora of Terra del Fuego, the Cape, and Australia, must be the remains,—need this conjecture raise difficulties at a moment when from their fossil remains Heer restores to our sight the ancient forests of Smith's Sound and Spitzbergen?"

Having ventured to reconstruct the southern continent, with its strange fauna, of which the remains are so widely dispersed, Rütimeyer casts about for more specific evidence in favour of the hypothesis to which the course of the world's formation everywhere gives rise, that fresh-water animals and likewise terrestrial animals came up from the sea. Hence the notably small division of sirenoid fish (Lepidosiren, Protopterus), which breathe air during the dry season of the year, must not be considered reptiles adapting themselves to aquatic life, but the reverse. The organ which in fish served as a hydrostatic apparatus, the swim bladders, becomes in them the lung. Thus we must go back from terrestrial to aquatic tortoises, and from them to those denizens of the sea which are allied to the Enaliosaurians, so frequent in the Jurassic strata. The evolutionary and biographical history of the land crabs shows us in the plainest manner how the inhabitant of the sea becomes a terrestrial animal; a special problem which, as we have already mentioned, Fritz Müller has completely solved and capitalized, in his essay, " for Darwin." Of the sirens, commonly but erroneously reckoned among the Cetacea, and of which the majority prefer remaining at the mouths of large estuaries, one entire species has penetrated into the great inland lakes of Africa; and certain species of salmon as well as the sturgeons, which alternate periodically between salt and fresh water, are in the phase of gradually forsaking ocean life. From my special experience, I may add that the brackish-water sponges are certainly dependent on the marine families, and that the fresh-water species unmistakably point to these brackish forms.

If in all these cases we are dealing with gradual transformation, and more or less voluntary adaptation, there is no lack of conspicuous instances of forcible and almost sudden severance; of upheavals by which former sections of the ocean became inland seas. What were the modifications undergone by the fish and crabs secluded with them, is shown by the fine observations of Lovén on the animals of Lakes Wener and Wetter, and of Malmgren on those of Ladoga. The latter brings evidence that the salmon-trout of the Alps (Salmo salvelinus) is derived from the Polar Sea, and is own brother to the Scandinavian Salmo alpinus.

Rütimeyer pronounces the opinion that by more minutely tracing the relations of the fresh-water fauna to those of the denizens of the ocean, the cosmopolitanism of fresh-water animals will be explained, as well as the relation of antarctic to arctic life. For the present, however, these two great animal groups, as regards the higher, warm-blooded classes, are somewhat sharply contrasted. It is only from scanty remains that we know that so early as the Jurassic era, the northern hemisphere was peopled by Marsupials, but, it is evident, not densely. We must suppose that, retaining their character, the Marsupials of the southern continent tested and proved their powers of adaptation, whereas on the other side of the equator a race of mammals of completely different cast proceeded from them. This is the race which still characterizes the whole surface of the earth from the north to the point of contact with the more stable remnants of antarctic life. While with reference to their origin we can appeal only to reason and inference, the historical connection between the mammalia now peopling the Old and the greater part of the New World, and their predecessors up to the most ancient Tertiary periods, is manifest to our eyes.

The remains of the earliest mammals here to be considered, are found in the Eocene deposits of Switzerland, and in corresponding strata in France and the south of England. From the southern edge of the Jurassic plateau, neither the Alps nor any other land was visible, and the ocean which washed its shores has been traced as far as China. The mammalia of this period, as far as they are known, amount, according to the synopsis made by Rütimeyer in 1867, to at least 70 species. The majority are ungulate, therefore Graminivora ; of these, by far the greater number Pachydermata. Now, when the entire world scarcely maintains so many Pachyderms, this ratio is quite disproportionate. In Europe, the pig alone represents this division, and Ruminants everywhere predominate. In its present animal population, Africa might be approximately compared to Eocene Europe. But as to these Ungulates must be added a large number of Carnivora, resembling the Viverrida (polecats, martens, &c.) and hyenas, and as viverridæ exist in Africa as well as in Asia, and as, moreover, the musk ruminants represented in this primitive fauna are now likewise Asiatic and African, and, finally, as the French opossums of those ages still live in Central and South America, "we gain an impression that the most ancient Tertiary fauna of Europe is the source of a truly continental animal society now represented in the tropical zone of both worlds, but most emphatically in Africa."

Far more heterogeneous is the picture of the higher animal life of the middle and more recent Tertiary periods which we reconstruct from the numerous and in parts highly prolific repositories of these remains. To draw narrower limits within these periods is impracticable; from place to place, from stratum to stratum, there is coherence; nowhere does a species appear that might not be derived from another; and our authority says that anatomy, morphology, palæontology, and geographical distribution, seemed to impress no doctrine upon him with such energy and pertinacity as that separate species of a genus, species without any historical and therefore without any previous local link to any original stock, do not exist." The most celebrated repository of Tertiary mammals is Pikermi, a short distance from Athens, an accumulation of skeletons complete and in fragments, which pre-supposes a profusion of animals, of which at any rate the most densely inhabited regions of Africa may, according to Livingstone's descriptions, give us an idea.

Again the Carnivora give way to the Graminivora,

though the feline beasts of prey make themselves conspicuous; and among the great Tertiary beasts of prey are some which have a range as great as the tiger of the present age. The territory of the extinct sabre-toothed tiger (Machairodus) at that time extended over a great part of America and Europe. Let us also mention that the canine animals appear somewhat later, and that the bears are of still more recent origin. At this period the most abundant material still favours the ungulates. Cloven feet still preponderate. Pigs and musk-animals are the most constant. But the tapir, in shape like the older forms, is now joined by the rhinoceros, the true horses, and the elephants. If the origin of the rhinoceros is somewhat obscure, the extraction of the mastodon, the older form of the elephant, is hitherto quite unknown.[73] And yet though we search in vain through the known mammalian fauna of the Eocene period for the most nearly allied parent forms, there are numerous tokens that even in Europe and Asia, "most of the Eocene must be regarded as the true root forms of the Miocene genera." (R.) This is shown by the discoveries at Nebraska in North America, where important genera, which, like the Palæotherium, disappeared from the Old World in the Eocene period, took refuge in company with newer genera. We likewise find there, intermediate forms between the lama and the camel, which in this case alone gives its true significance to the once unmeaning word, vicarious genera. At Nebraska we moreover find the triple-hoofed horse (Anchitherium), and we hence know the origin of the single-hoofed horse of the Old and New Worlds.

What has happened in the Old World since that age

is confined to the extinction of many Pachydermata, a displacement of the rhinoceros, elephant, tapir, and hippopotamus, and an extremely abundant development of the true ruminants and the cattle which proceeded from them with an exaggerated form of head. Bears and canine species occupy the territory where viverridæ and hyenas once predominated; but as "numerous locally and historically limited species, a large number—among the smaller fauna a majority—of Miocene races remain in possession of the ancient and probably constantly increasing habitat." (R.) "In this gradual change of things, no one will be able to discern aught but phenomena of the same order of which we are still the witnesses." (R.)

How circumstances occurred in America has been described in a masterly style by Rütimeyer as follows: "America affords a basis for the distribution of animals completely different from that of the Old World. In the latter, ridges, open only in places, divide the entire continent into mountainous zones, and correspond to the distribution of temperature. Thus in a twofold manner they prescribe a definite range east and west to the extension of animals; while a migration from north to south is impeded less by the height of the mountains than that on their summit the north comes into contact with the scorching south. Behind this wall, moreover, in the expanse from the Caspian Sea to China, there is a zone of steppes and deserts which fences in the animals more effectually than the mountain chains. In America, not beasts of prey alone, but graminivora also, may advance without hindrance from the regions of the lichen on the Mackenzie River, through the pine forests of Lake Superior, to the land of the magnolia in Mexico; 40°—50° of latitude separate the extremes which meet in the Himalayas, and the vast plains and huge river systems seem almost to solicit immigration. The accordance of the whole faunas of Mexico and Guiana, moreover, shows how little the isthmus of Panama checks the advance to South America, where again one mighty fluvial system trenches upon the other without any lofty partitions; nor is there any arid desert in the whole extent from the Canadian seas to Patagonia.

"We shall probably not be wrong in ascribing the remarkable extension of fossil and present mammals of America in a great measure to this circumstance. As we have seen, the Miocene fauna of Nebraska is the offspring of the Eocene fauna of the Old World. The Pliocene animals of Niobrara, which are buried in the same district as Nebraska, but on more recent arenaceous strata, still further corroborate this statement: elephants, tapirs, and many species of horses, scarcely differ from those of the Old World; the pigs, judging by their dentition, are descendants of European miocene Palæochœridæ. The ruminants are represented by the same genera, and partially by the same species, as in the analogous strata of Europe, as deer, sheep and buffaloes; neither do the carnivora or the minute animal life offer an exception. Many genera of an entirely Old-World cast have in the lapse of time penetrated far into South America, and there died out shortly before the arrival of man, or perhaps by his co-operation, as was the case with the two species of mammoth of the Cordilleras and the South American horse, whose present successors reached this insular continent by a far shorter road. Even a species of antelope and two other horned ruminants (Leptotherium) found their way to Brazil. Two sorts of tapir, of which the dentition, even in Cuvier's eyes, is scarcely distinguishable from the Indian species; two species of pigs, still bearing in their milk-teeth unmistakable characters of their aboriginal form; and a number of deer, besides the lamas, a later and originally American offshoot of the Eocene Anoplotheria—are one and all living remnants of this ancient colony from the East, which did not reach its dwelling-place without copious losses on its long pilgrimage. It can scarcely be doubted that many of the beasts of prey which in the Diluvium of South America retained their family character more than they do now, must have arrived there in the same manner. Let us now remember that even the Eocene Cænopithecus of Egerkingen distinctly pointed to the present apes of America, and that the Didelphidæ (Opossums) lie buried in the same European soils. It might almost appear that it was pre-eminently the division of arboreal quadrumana which, with the opossums, domesticated itself in the vast forests of their new abode, and, receiving a fresh impulse, gave rise to a multitude of special forms, without however having, even in the present times, reached the pitch of development attained by their cousins who had remained behind in the Old World.

"We may now appropriately return to our previous remark that this migration of animals did not find the south of the New World destitute of mammals, but rather already occupied by the toothless representatives of antarctic, or at least of southern animal life. The diluvial fauna of South America collected by Lund, Castlenau, and Weddell, from the Brazilian caves, and the alluvium of the Pampas, among the 118 species cited, actually includes, in addition to those already mentioned, as being of probably Old-World pedigree, no less than 35 species of Edentata, and these animals of considerable bulk. Not reckoning the 36 rodents and bats, and the smaller fauna in general, they constitute nearly half of the larger diluvial animals of South America. The assemblage of Edentata previously settled in these regions thus held their own against the invasion from the north.

It is comprehensible that the same external causes which led the march of the children of the north constantly further, may likewise have invited the members of the antarctic fauna to extend themselves northwards. As we even now encounter the incongruous forms of the sloth, the armadillo, and the ant-eater in Guatemala and Mexico, in the midst of a fauna in great part consisting of races still represented in Europe, we also find, even in diluvial eras, gigantic sloths and armadillos ranging far into the north. Megalonyx Jeffersoni, and Mylodon Harlemi, sentries of South American origin thrown out as far as Kentucky and Missouri, are a phenomenon as heterogeneous in the land of the bison and the deer, as is the mastodon in the Andes of New Granada and Bolivia. Over the whole enormous extent of both portions of the New Continent, the mixture and interpenetration of two mammalian groups of completely diverse families, constitutes the most conspicuous feature of its fauna; and it is significant that each group increases in the abundance of its representatives and in the originality of their appearance as we approach its point of derivation."

Hence, on both sides of the ocean, north of the very sinuous boundary of the antarctic or southern fauna, we find ourselves still in the midst of the diluvial animal world, which extended itself, by a bridge in the vicinity of the North Pole, from the old continents to the mainland of America, and there for a longer period retained its ancient appearance in the mastodons and horses.

There, as well as here, the present order of things—the cantonment of animals—has been in many ways determined and modified by mighty glacifications and prolonged periods of refrigeration. Hence the accordance of so many plants of the extreme north with Alpine plants after the Eocene vegetation had made its entry from the east. Since that age, the reindeer has been forced back to the north, and the musk ox has been expelled and exterminated from the Old World. The elephants, fleeing before the ice, have not returned; and the mammoth, immigrating with a rhinoceros from the north-east, has been destroyed with his associate. Others of his comrades, such as the primæval ox, died out only a few centuries ago as wild cattle; others, like the buffalo and the beaver, are nearly extinct as denizens of Europe; and others again, the deer and roe-deer, will perish with the forests and the game-laws. But of almost all the species of which we search for the extraction, Palæontology supplies us with the history and derivation; and in derivation we find the causes of geographical distribution sketched in vivid outlines.

XI.

The Pedigree of Vertebrate Animals.

THE final result towards which the doctrine of Descent directs its efforts, is the pedigree of organisms. To work it out is to collect the almost inconceivable profusion of facts accumulated in the course of about a century by descriptive botany and zoology, including comparative anatomy and the history of development, and to submit the existing special hypotheses to a minute scrutiny and renewed verification. We have therefore claimed in behalf of the doctrine of Derivation the privilege on which the progress of science generally relies—that of investigating according to determined points of view, and accepting probabilities as truth in the garb of scientific conjecture or hypothesis. It is manifest that when the doctrine of Descent first made its appearance with the arguments proposed by Darwin, it was only possible to indicate the most general outlines of this great pedigree, which it was the special task of the new direction of science to demonstrate in all its details. But however and wherever specific research was attempted, either the results contributed the form of some part of the great pedigree, or there was, from the first, reason to pre-suppose certain kinships, and the conjecture was tested. The further an inquirer has carried his survey of the conditions of organization in any of the larger groups, the less will he be able to divest himself of the genealogical idea in his every act and thought.

All this is so self-evident, that one would scarcely suppose that the use of this method could have been made a subject of reproach to the doctrine of Descent. Nevertheless, it frequently occurs, and the champions of the doctrine of Descent are blamed for often speaking of mere probabilities, forgetting that even in cases in which the probability ultimately proves false, the refuted hypothesis has led to progress. Of this the science of language has recently borne testimony. It is well known that linguistic comparison within the family of Indo-Germanic tongues suggested the reconstruction of the primitive language which formed their common basis. Johannes Schmidt[74] now proves that the fundamental forms disclosed may have originated at widely different periods, and hence that the primitive language, regarded as a whole, is a scientific fiction. Nevertheless, inquiry was essentially facilitated by this fiction, and with it was intimately connected the formation of a pedigree of the Indo-Germanic linguistic family, as a hypothesis supported by many indications. A bifurcation was assumed into a South European language, with Greek, Italian, and Celtic ramifications, and another language, from a second division of which proceeded the fundamental language of North Europe and the Aryan fundamental language. Although Johannes Schmidt has demonstrated that this pedigree is false, as the existence of Slavotic shows the impossibility of the first division assumed, the value of the hypothesis is undiminished. It was the road to truth.

In our science Haeckel has made the most extensive use of the right of devising hypothetical pedigrees as landmarks for research. It matters nothing that he has repeatedly been obliged to correct himself, or that others have frequently corrected him ; the influence of these pedigrees on the progress of the zoology of Descent is manifest to all who survey the field of science, not to mention that in the last ten years a series of researches have conclusively fixed their results in good pedigrees. As we propose to give merely an introduction to the doctrine of Descent, we shall content ourselves with showing how the system or the pedigree is constituted in its application to the single group of the Vertebrata.

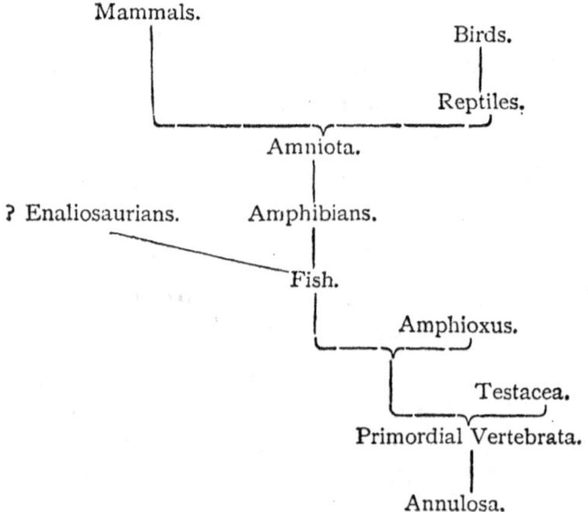

As we have seen above, the most important indications of the pedigree of the species are contained in the evolutionary history of the individual. Only, if all vertebrate animals testified their family connection by agreeing *inter se* in the distribution of the germ as well as in the fundamentally important organs, the spinal cord and the vertebral column, this token of their descent from inferior animals, which is unconditionally demanded by the theory, seemed to be entirely wanting. In other words, it seemed that in all vertebrate animals the memory of their original derivation had been obliterated by curtailed development (comp. p. 211). Thus the case remained until Kowalewsky a few years ago studied the development of the lancelet (Amphioxus), the lowest vertebrate animal known, and showed that in this creature the typical phenomena of vertebrate development

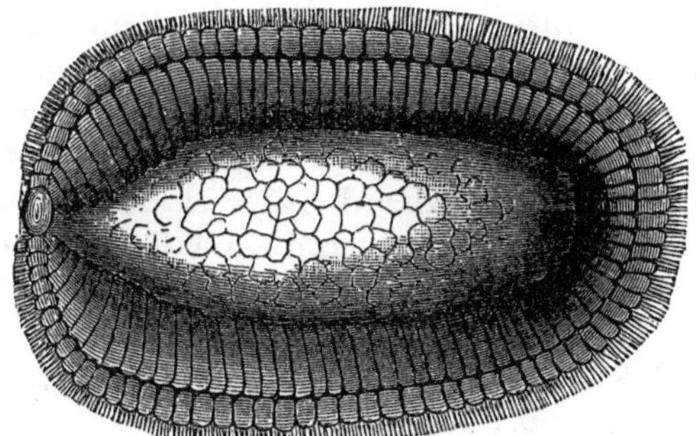

FIG. 22. Larva of the Lancelet after Kowalewsky.

are preceded by the phases required by the theory. We have already made acquaintance with this form of development (p. 51, &c.), and we here again point out its profound significance. It is only when the Amphioxus has passed through the phase of the vibrating, sac-like gastrula larva that the future dorsal side becomes flattened, and the protuberances arise, which shortly after close into the sheath of the spinal marrow, while underneath originates this important cellular column, the chorda dorsalis, or notochord. With this the lancelet becomes a vertebrate animal, and the preceding phases do not (according to the view at one time inculcated by C. E. v. Baer respecting such phenomena) recall the inferior and undeveloped in general by the absence of differentiation, but they agree in genesis and distribution, in the differentiation of their cellular layers, and in their totality, with the Gastrula phases of invertebrate animals.

We are therefore fully justified in regarding these first incidents in the evolution of the Amphioxus as a reminiscence of the roots of the pedigree of the Vertebrata; and this direct indication of the descent of vertebrate from invertebrate animals is supported by a second and no less important discovery by the Russian naturalist. It is, that during their development a number of the Testacea of the division of the Ascidians temporarily possess a spinal cord, and the rudiments of a vertebral column. Kowalewsky's researches have been ratified on all essential points and in many ways extended by Kupfer, and the facts which interest us may be explained by the diagram, Fig. 23, representing the point of the larva of an Ascidian in a somewhat advanced stage. The bulk of the Ascidian larva consists of a body of which our figure shows the whole, and a rudder-like tail. The appendages projecting from the body on the right are organs of adhesion, by means of which the larva fixes itself for its definitive transformation. At *o* the orifice of the mouth is formed; *d* developes into the branchial cavities and the intestinal

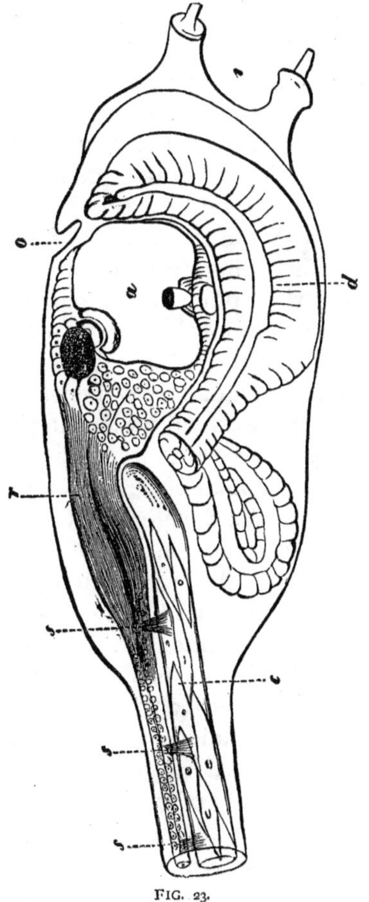

FIG. 23.

canal, and we will incidentally remark that in the lancelet also, the anterior end of the primitive intestine becomes the branchial cavity. But with reference to the vertebrate animals, the most important parts of the Ascidian larva are the following. It possesses a true spinal cord with a vesicularly expanded brain (*ra*). The distribution and position of this organ agrees accurately with the corresponding parts of the vertebrate animal, and Kupfer has even discerned the rudiments of nerves (*s s s*), which, if the observation is confirmed, will still more incontrovertibly establish the homology of the organ in question with the spinal cord of the Vertebrata and the nerves proceeding from it in pairs. But we know that it is not the spinal cord alone, but its combination with the vertebral column which constitutes the characteristic feature of the vertebrate animal. This vertebral column the Ascidian larva likewise possesses (*c*) in the form of the noto-chord, and, as in the vertebrate animal, this embryonic vertebral column lies between the intestine and the spinal cord. So far goes the accordance; henceforth, the development of this part, so important to the vertebrate animal, becomes retrogressive in the Ascidian. The rudder-like tail, with the spinal cord contained in it, and the noto-chord, are cast off when the animal becomes fixed; the larval brain which promised so well, shrinks into an insignificant nervous ganglion, and the complete animal gives no cause for suspecting its analogy with the Vertebrata.

These laborious observations prove that the Vertebrata are not the sole proprietors of the spinal cord and vertebral column, but received these organs as a heritage from lower grades of organization as their progenitors. It does not occur to the Darwinists to regard man as the direct offspring of the present apes; neither do they infer from these observations on the Ascidian larva that vertebrate animals are descended from the Ascidians. Their accordance much rather forces us to assume an unknown primordial vertebrate family, springing from some branch of the heterogeneous division of the Annulosa. From these diverged on one side the Testacea, who might perhaps be called mischanced vertebrata, and on the other the true vertebrate animals.[75]

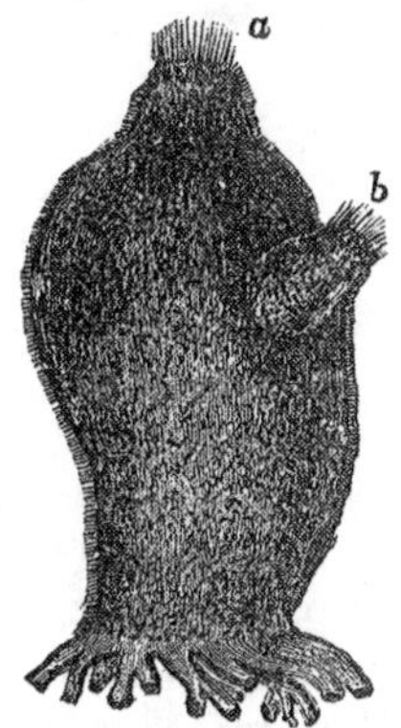

FIG. 24. Full-grown Ascidian.

The Amphioxus which lives in the sand in shallow places on various coasts, and is daily caught by thousands at Messina for example, is five or six centimetres in length, and is compressed after the manner of a fish, pointed at both ends, and semi-transparent whilst alive. It possesses no trace of limbs, at the posterior end only a pair of minute membranous margins, the indication of dorsal and caudal fins, and is so simple in its internal structure that it is usually, though inaccurately, termed a fish. Its skeleton is limited to the noto-chord, and some minute cartilaginous rods at the mouth and gills. It has no brain, and, except a small ciliated sac, perhaps to be interpreted as an olfactory organ, no sensory apparatus; the heart is tubular. And thus between the lancelet and other true fishes there exists so wide a difference that the possibility remains open that the fishes passed through some other course of development than phases like that of the Amphioxus.

Our knowledge of the genealogy of the fishes may be laid down in the following diagram :—

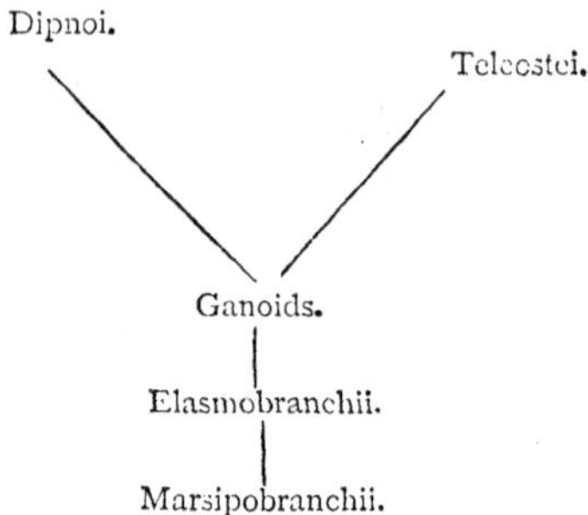

The Marsipobranchii (Cyclostomi), it is true, exhibit important peculiarities, such as deficiency of limbs, entire absence of bony plates or scales on the integument ; but the brain, heart, and vertebral column (which, although persistently cartilaginous, is far superior to that of the Amphioxus), show their direct coherence with the fishes. Fossil remains of these animals, universally known in the genus lamprey (Petromyzon), are not forthcoming, and, at the most, only their horny teeth could have been preserved.

After these manifest gaps in our knowledge, the succeeding orders of fishes present themselves in a connection all the more conspicuous. The starting-point is formed by the Elasmobranchii, to which belong the true chimeras, sharks, and rays. Brain and gills testify their kindred with the Marsipobranchii. In the construction of the cranium, facial bones, pectoral and pelvic arches, and the anterior extremities, heart and intestine, they exhibit forms to which, as Gegenbaur has shown in his well-known observations, the homologous parts of the Ganoids are related either as progressive developments or as reductions. Huxley has also prepared the way for a correct apprehension of these relations. To be fully convinced of this, detailed study is certainly requisite; for in its absence it is impossible to imagine, how in the Elasmobranchii the true branchial apparatus is wanting, and how the cartilaginous arch, which, in them, replaces the gills, is applied in the Ganoids, partly as the palate, and partly as the attachment for the true lower jaw, while the internal gills of the former, become the external gills of the latter; how in the skeleton of the anterior extremities, a gradual simplification may be exhibited, step by step, from the sharks and rays to the Ganoids, and especially the sturgeon,—a process of which the two extremes are reached in the Teleostei on the one side and the higher Vertebrata on the other—in the latter in the multiform perfection of the arm and hand.

Of the Ganoids only scattered remnants survive, the sturgeon family and some few American and African genera, of which, as Rütimeyer says, a flight into fresh water has been the salvation. They just suffice to explain the relation of this once extraordinarily extensive group, to the Elasmobranchii as well as the Teleostei.

In the Teleostei, the metamorphosis of the organization of the Elasmobranchii initiated in the Ganoids, is carried yet further. It is only with great qualification that they can be termed "more highly developed," in the skeleton perhaps, to which older zoologists attributed too much importance. Brain, heart, the form of the extremities, and the reproductive system, are indeed distinct developments which, in combination with the external shape and integuments, have exhibited great powers of adaptation, but have not proved capable of any further development. Comparative anatomy has vainly spent much labour in attempting to trace the condition of the higher animals from the special organization of the Teleostei, or to explain the peculiarities of the Teleostei from above downwards. It was labour lost, for the solution is to be reached only by the method indicated in the derivation of the Teleostei, through the Ganoids, from the shark-like fishes.

Hence, at the present period, a development is concluded with the Teleostei, and we must look to another grade for the transition from the fishes to the amphibians. We find one in the order of the mud-fishes (Dipnoi), scantily represented by only few species (Lepidosiren Protopterus). These fish-like animals, living in American and African rivers which dry up in the hot season of the year, are fish by right of their skeleton and scales, and some other characteristics; the skull, however, almost resembles that of an amphibian, and they also provisionally use their swim-bladders as lungs; and by thus breathing alternately water and air, they set before us the transition of the gill-breathing larvæ of the amphibians to the phase of air-breathing. Of the true fishes at the present time, they most nearly approach the family of the Crossopterygii, represented by the African Polypterus; and the discovery of a very remarkable Australian fish, the Ceratodus, confirms this affinity.

Through forms thus resembling the Dipnoi, the advance from the fishes to the amphibians was probably accomplished. But, as a scientific friend, profoundly versed in the history of development, has pointed out to me,—supporting his remark on the comparison of the respiratory organs of the Marsipobranchii with those of the amphibians,—it is possible that frogs and salamanders may be directly descended from beings closely analogous to the division of the Marsipobranchii termed Myxine. It is to be hoped that this highly interesting observation may soon be made public. We gather from the general Ontogenesis of the amphibians, that the tailed forms are the most ancient. This is also the case with the oldest amphibian-like animals, the Labyrinthodonts. From their remains (Archegosaurus and others), chiefly contained in the Carboniferous formation, we have learnt that they had incomplete limbs or none, that their ventral side was partially provided with bony plates, the vertebral column fish-like, and that their skull, with some of the characters of the present amphibians, combined others which remind us partly of certain bony Ganoids, and partly of the reptiles which subsequently appeared. Now if in the singularly elongated snake-like Cœcilia, which is however without tail or limbs, some peculiarities of the skull of the Labyrinthodont appear again, we must own our utter ignorance as to the actual progenitors of this, as well as of the two other living orders of the Cœcilia and the Batrachians. Here, therefore, we are, as we have said, thrown entirely on the evolutionary history of the individual. By what right we may frame a picture with great probability approaching the truth, the reader may have gathered from our previous chapters.

Among the tailed amphibians, it is not only in Ontogenesis that we see the passage from gill to lung-breathing; the systematic series from the proteus to the triton and the salamander, likewise exhibits this physiological ascent, linked with various morphological transformations, which may similarly be shown between the ancient and modern specimens of the Labyrinthodonts. The Batrachians, indeed, rise higher in development than the Cœcilia; but, as the friend above mentioned informs me, they more nearly approach the Myxine in the construction of the internal gills of their larva. We shall obtain a general view of the reptiles by means of the appended diagram, in which we shall avoid any minute systematic designations.

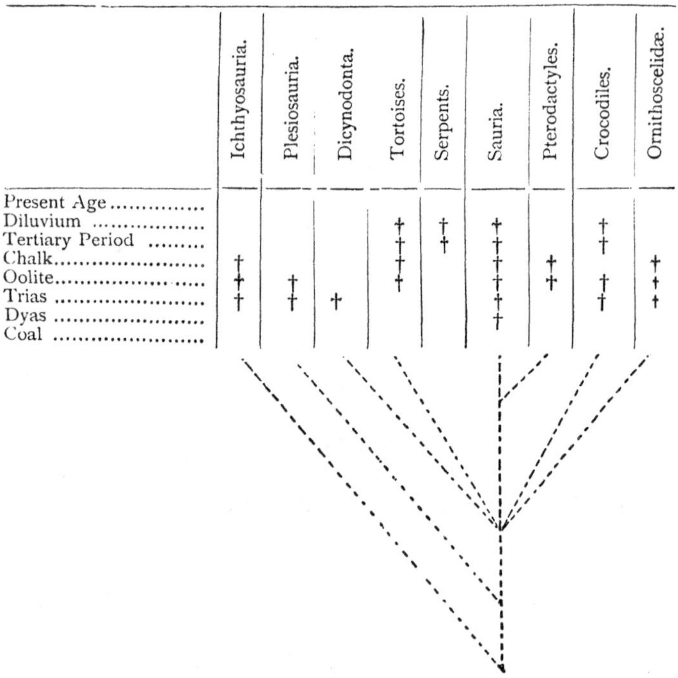

	Ichthyosauria.	Plesiosauria.	Dicynodonta.	Tortoises.	Serpents.	Sauria.	Pterodactyles.	Crocodiles.	Ornithoscelidæ.
Present Age									
Diluvium				†	†	†		†	
Tertiary Period				†	†	†		†	
Chalk	†			†		†	†		†
Oolite	†	†		†		†	†	†	†
Trias	†	†	†			†		†	†
Dyas						†			
Coal									

The class presents a very comprehensive picture, although only four orders now exist, of which two, the lizards and the snakes, are scarcely to be separated. That the snakes, which first appear in the Tertiary period, are a direct offshoot from the lizards, is reduced to a certainty by comparative anatomy and the history of development. In the various families of lizards we see the absence of feet occurring in conjunction with the elongation of the body and the multiplication of the vertebræ; and the modifications peculiar to the skull of the "true" snakes are likewise represented in the systematic series in every gradation, beginning with the skull of the true lizard. We cannot specify the fossil genera in which the transformation was initiated; but in this case a doubt would be only a capricious denial. It is otherwise with the remaining orders, which in the beginnings, hitherto accessible to us, exhibit diversities so decidedly marked, that in none has it been possible to trace a direct descent from any known member of another. Prof. Huxley, a great authority on the anatomy of these animals, says on this subject as follows:—

"If we ask, in what manner the earliest representatives of these orders are distinguished from their living or latest known representatives, we shall find, in all cases, that the amount of difference in itself is remarkably small in comparison with the length of time during which the order has existed. So far as I know, there is no fact to show that the later Plesiosauria, or Ichthyosauria, exhibit an advance upon the earlier members of the group. It is not clear that the Dinosauria of the wealden and of the Cretaceous formations are more highly organized than those of the Trias; and even where a differentiation of structure is to be observed, as in the Lacertilia, or Crocodilia, it goes no further than a modification of the form of the articular surfaces of the vertebræ, or of the degree to which the internal nasal apertures are surrounded by bone. The osteological differences, which alone are exhibited by fossil remains, have doubtless been accompanied by many changes in the organization of the destructible parts of the body; but everything tends to show that the amount of change in the organization of reptiles since their first known appearance upon the earth, is not great in itself; and is wholly insignificant, if we take into consideration the lapse of time, and the changes of the surface of the globe, which are represented by the Mesozoic and Tertiary formations.

"From the point of view of the evolution hypothesis, it is necessary to suppose that the Reptilia have all sprung from a common stock, and I see no justification for the supposition that the rapidity of their divergence from this stock was greater before the epoch of the Trias than it has been since. Consequently, seeing that the approximation of the oldest known representatives of the different orders is so slight, reptiles must have lived before the Trias for a length of time, compared with which that which has elapsed from the Triassic epoch until now is small—in other words, the commencement of the existence of reptiles must be sought in a remote palæozoic epoch."

Comparison thus points us back to ages which afford no record of the actual derivation of this class. Even the Ichthyosauria and Plesiosauria, so frequently mentioned in conjunction, deviate widely from one another in very essential characters, which refer their supposed common origin to a remote period. We will mention only the fin-like extremities of the former, which are of an obviously piscine type. We are thus thrown back vaguely on such mixed forms as may have been analogous to the Labyrinthodont; nay, the question arises whether the Ichthyosauria alone, or perhaps the Plesiosauria with them, did not diverge from the fishes independently of the other branches of the reptile family; an eventuality which is taken into account in the pedigree at p. 250. A certain resemblance with the skull of the tortoises (Chelonia) is exhibited by that of the Dicynodonta. In them also the jaws, as appears from their shape, were manifestly cased in horny sheaths; but at the same time the upper jaw contained two huge tusks, and it is scarcely possible to imagine a direct transition from the Dicynodonta, appearing in the Trias, to the more recent tortoise. In some particulars of the skull, as well as in the situation of the posterior nasal apertures, the forms of older crocodiles exhibit an affinity with the lizards, from the older and unknown forms of which they probably branched off. The winged saurians, or Pterodactyles, may also be a branch of the lizards. They have gained by adaptation several characters, such as the shape and lightness of head, the length, slenderness, and pneumatic character of the tubular bones, which they share with the birds. But it is not in them, but in the division comprising several families which Huxley terms Ornithoscelidæ, or reptiles with the legs of a bird, that we must look for the actual progenitors of the birds. For among them one of the most important characters of the birds is, in some genera, in course of preparation, so that in the full-grown animal its origin may still be recognized; in others, as the genus Campsognathus, it is accomplished. We allude to the peculiarity already discussed in p. 10, that the upper portion of the tarsus is anchylosed with the tibia, the lower with the metatarsus, and that the ankle-joint is hence inserted into the tarsus.

All existing reptiles are sharply distinguished from the Amphibians and Fishes by several phenomena accompanying their development. They possess two organs enveloping the embryo; the amnion, which is essentially a protecting sheath, and the allantois, by which the fœtal circulation, nutrition, and respiration is regulated and carried on. In the Batrachians we find indications at least of the allantois, and must suppose that the greater part of the fossil reptiles had already adopted this advance in general organization. It implies an advance, inasmuch as animals developed by the aid of the amnion and allantois make further progress during the embryonic phase than is the case with the inferior Vertebrata, and that theyhence leave the egg with greater powers of resistance. We must ascribe the adoption of the amnion and allantois to remote periods of amphibian and reptile development, for the additional reason that the possession of their embryonic sheaths and organs is shared by the birds which are descended from true reptiles, and by the mammals which cannot be descended from true reptiles.

The birds are, anatomically, so closely allied to the reptiles, that Huxley, who has carried out the comparison most rigorously, has joined the two classes into a greater systematic unit, under the name of Sauropsida, or lizard-like animals. The scale of a lizard and a feather seem to be totally different things; but in their first rudiments they are completely identical, and the feather has a far greater analogy with the scale, than with the hair. The plumage, which seems to impress a specific character upon the bird, is therefore to be traced from the formation of scales. Of the internal soft organs, we will only remark upon the heart and lungs. All the older geologists placed the heart of the bird on the same level with that of the mammal and of man; in its specific arrangements, however, it is only to be interpreted by the heart of the reptile, and the wind-pipe is not ramified as in the mammal. That the reptiles exhibit a gradual transition to the leg of the bird, has been repeatedly pointed out. The pelvis of the bird, which is remarkable for the length of the pubis and ischium, and is open in front, likewise represents only a slight advance in development upon the pelvic structure already shown in several of the Ornithoscelidæ. Thus Huxley says with reference to the ischium of the Hypsilophodæ, that "the remarkable slenderness and prolongation of the ischium give it a wonderfully ornithic character." Finally, in the skull, peculiarities possessed by the bird in contrast with the mammal, such as the simple condyle of the occiput, the quadrate bone, the cochlea of the auditory labyrinth, the composition of the lower jaw, its articulation with the skull by the intervention of the quadrate bone, &c., are not specific characters of the bird alone, but of reptiles in general. This similarity of type in reptiles and in birds is perfectly manifest from the comparison of living birds with living reptiles. But the proof that the bird is derived from the reptile is rendered unimpeachable by the discoveries, scanty as they are, of fossil intermediate forms. The pelvis and leg of the Ornithoscelidæ have already been discussed. But in the slates of Solnhofen we have moreover become acquainted with the Archæopteryx, a bird unfortunately mutilated and in many ways damaged by pressure (Fig. 25, impression of the tail of the Archæopteryx Macrurus, Ow), but exhibiting a very valuable and interesting intermediate stage between the tail of a reptile and a bird. Among existing birds, the Nandu, or American ostrich (Rhea), alone possesses numerous separate caudal vertebræ; but the tail of this bird projects so little, that it in no way recalls the tail of a lizard. Now the Archæopteryx exhibits a long tail, bordered by two rows of stiff feathers, of which the impression remains in extraordinary preservation. The skull of this valuable specimen, now in the British Museum, is so much injured, that no idea can be framed of its construction. It is impossible to decide whether the jaws bore teeth. The example of the tortoises shows that within the reptile type the formation of teeth was replaced by horny sheaths, without a correlated development of the power of flight; the Pterodactyles, on the other hand, combine with the power of flight a light head, provided nevertheless with numerous teeth.

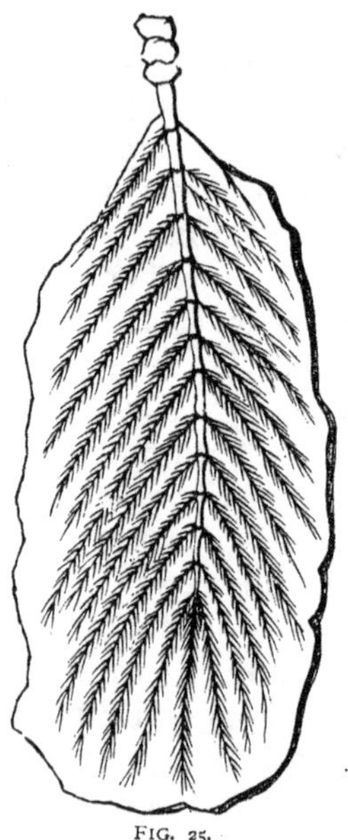
FIG. 25.

The obscurity which surrounded these parts of the old antediluvian birds has been cleared up by a discovery by the American naturalist, Marsh. He found in the upper Chalk of Kansas the remains of two genera of birds, which by their bi-concave vertebræ remind us of the characteristics of the ancient reptiles, and by this alone present extremely valuable intermediate stages, but which, moreover, bore teeth in both jaws. These teeth are small and sharp, and were so numerous that in the lower jaw of the animal named Ichthyornis dispar, twenty might be counted on each side.

Thus we are now quite clear as to the kinship of the bird. It is a reptile adapted to aerial life, and those birds which we see more estranged from flight have acquired the characters correlated with more or less incapacity for flight only by means of retrogression. It fares the worse with the internal arrangement of this class of animals. Partly from their geographical distribution, partly from anatomical indications, especially of the skull, it may be inferred that the ostrich-like birds are not, in virtue of their strength of leg and adeptness in running, the youngest members of their class and the most nearly allied to the mammals, but that they are the oldest of those now living. The nature of the imperfection of their wings shows, as we have said, that they are in a state of arrest or retrogression. Beyond this general experience it is impossible to go. If we contemplate the bird as a flying animal, those of course rank highest which have learnt to fly the best. This palm avowedly accrues to the birds of prey as a whole, although other orders are not deficient in pre-eminent flyers. Brehm and others hold the parrots, because of their docility, to be the highest birds. But all this is arbitrary, and can only accidentally correspond with the true and unknown ramification of the ornithic branch in the pedigree of the Vertebrata.

The most ancient known remains of the Mammalia are found in the Trias. They occur somewhat more frequently in the central Mesozoic strata, and they all belong to Marsupial animals. Now as Marsupials, in comparison with the inferior classes of vertebrate animals from which they must be derived, are very highly developed, and as in the Monotremata (Duck-mole, Ornithorhyncus, and Porcupine ant-eater, Echidna,) we possess mammals which are manifestly far beneath the Marsupials, we are referred entirely to conjecture and inference for the origin of the mammals. These point to amphibian-like beings, in which certain peculiarities of the mammalian skull, such as the double condyle of the occiput, were prefigured, and which by the formation of the amnios and allantois approached the true reptiles. These progenitors of the Mammalia are not, however, represented in any order of reptiles or amphibians now extant. The pedigree (p. 269) in which we have grouped the more accurately known fossil Mammalia with those now living, contains considerable gaps, and rests in a great measure on hypothesis, but it gives, nevertheless, with approximate probability a correct representation of the consanguinity of the orders, and in comparison with the system as it was constructed in the school-books prior to the revival of the doctrine of Descent, it must be esteemed a great and suggestive advance.

Man
Monkeys
Insectivora
Bats
Rodents
Lemurs
Beasts of Prey
Seals
Whales
(Primordial Beasts of Prey)
(Placental Mammals)
Marsupials
Monotremata.
Elephants
?
Hyrax
Sirens
Ungulates
(Primordial Ungulates)
Edentata

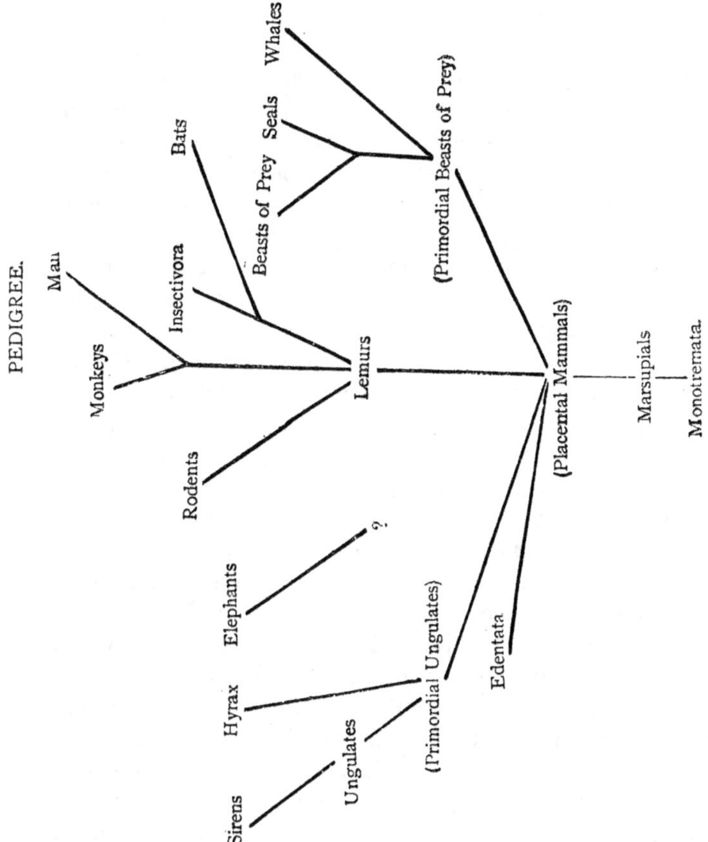

PEDIGREE.

As regards the structure of their skull, the constitution of the pectoral arch, and their persistence in the phase (embryonic in other mammals) in which the rectum and the urinary and genital ducts open into a single cloaca, the Monotremata (Ornithorhynchus, Echidna), limited to Australia and Tasmania, are the lowest members of their class, and must be considered as remnants of a division reaching from indeterminable past ages down to the present time. It may be presumed that the Marsupials were developed from an analogous grade. Their powers of adaptation have been chiefly testified in Australia, where the subdivisions of the order, usually designated as families, are, in dentition and habits of life, developed in a manner analogous to several of those orders which appear on the second great scene of mammalian development, namely, the Northern hemisphere.

Far advanced beyond the Monotremata as to skeleton, they remain on a low grade with respect to the reproductive system, and are implacental, like the Monotremata. That is to say, the embryonic blood-vessels do not enter into those close relations with the blood-vessels of the maternal ovary, by which the more perfect development of other mammals within the mother's womb is effected. This character and the correlative formation of the pouch in which to carry the immaturely born offspring, bind together the various families of Marsupials, which deviate from one another like other orders.

With the exception, therefore, of the two orders named, in all mammals the embryo is attached to the maternal organism by the so-called placenta. The blood-vessels of the developing offspring which reach the wall of the uterus by the intervention of the allantois, form coils and loops, between which grow similar offshoots and appendages of the blood-vessels of the ovary, so that through the walls of the contiguous blood-vessels an abundant exchange of fluids takes place between the two, and therewith a prolonged nutrition and a further and more complete development of the fœtus. The higher character of the placental mammals, usually plainly evinced by their anatomical relations, is thus based on the existence of the placental mass. All intermediate grades are, however, wanting which would entitle us to infer with certainty the direct transition from implacental to placental mammals. The Edentata, (Bruta), manifestly the lowest of placental mammals, are so devoid of any nearer morphological relations with the Marsupials, that we must needs be content to assume generally, on these indications, supported by geographical distribution and geology, that they represent a very ancient branch of the placental mammals. As we saw in the tenth chapter, they are scattered remnants which can only by compulsion be united into a single order. Sloths, armadilloes, ant-eaters, differ from one another at least as much as rodents, insectivora, and bats. The doctrine of Descent is not discredited because it is unable to account for these fragments of bygone animal life, but in the absence of data it is for the time in presence of an impossibility.

To ascertain the relationships of other orders, the modern systematizers, and also the supporters of the system of Descent, have thought fit to lay great stress on the presence or absence of the so-called decidua. This requires a short explanation. In many orders of mammals, the vascular processes and villi of the wall of the ovary become so closely connected with the fœtal portion of the placenta, that at birth the entire membranous coating of the ovary is detached and thrown out with it. In others, the vascular villi are not so closely adherent; they yield without important lacerations, and hence no deciduous membrane (Membrana decidua) is ejected. Now, as it appears to me, the specific conditions of the formation of the decidua have been far too little compared to justify our inferring any close affinity from the mere fact that portions of the coating of the ovary are lost in parturition. Much rather it must be unreservedly admitted that the formation of decidua might be occasioned by subordinate circumstances of the most varied kinds, and hence in orders only remotely allied, or allied merely as placental mammals. We therefore consider the decidua to be a subordinate systematic feature where anatomical and morphological reasons are opposed to it.

We go yet further. In the modern system the form of the placenta is likewise employed in the grouping of organisms. If among the Deciduata, lemurs, rodents, insectivora, bats and monkeys, are classed together as orders with discoidal placenta, this combination is certainly supported by a series of other reasons, and it is quite probable that within this group of orders the form of the placenta is due to homology, that is to Descent. But when beasts of prey, elephants and the Daman (Hyrax) are further cited as orders with zonary placenta, we find ourselves in the same position as when the decidua was reckoned decisive as to the closer affinity; and we are of opinion that the subordinate form of the placenta might similarly arise in different ways, just as it has been variously developed in the well-substantiated division of the Ungulata. To corroborate our view by example, we are certainly unable to make any positive statements as to the derivation of the Proboscidæ. It is, however, none the less certain that nothing positive is implied by the customary classification by reason of their zonary placenta. But we shall more nearly approach the truth if we place this branch of unknown origin typically nearer to the Ungulata than to the beasts of prey. If, moreover, as non-deciduate mammals, the Cetacea are held to be more closely allied to the Ungulata than to the Carnivora, which are deciduate,—in our eyes, this circumstance is not decisive, as more important reasons argue that the Cetacea were first developed from carnivorous genera.

In our exposition of the geographical distribution of animals, we derived instruction from Rütimeyer with

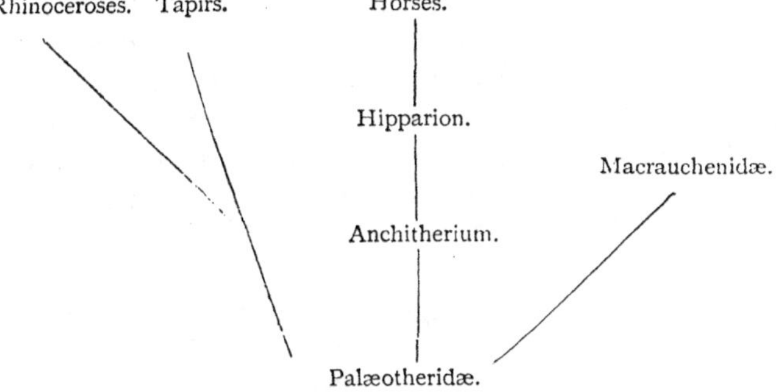

reference to the relationships of the Ungulata in particular. In no other division do we possess such abundant fossil material. In the older Tertiary strata we encounter the remains of two Ungulate families, the Palæotheridæ and the Anoplotheridæ, essentially distinguished from one another by their dentition, and forming the starting-points of the groups of Ungulates of which some now appear so greatly isolated. The root to which these two families lead back is unknown; on the other hand, partly from the direct comparison of these genera with the present Ungulata, partly from numerous intermediate links found in the Miocene, Pliocene, and Diluvium, it appears that, in the lapse of time, the separation which characterizes the present age was initiated, and the seeming isolation was produced by the extinction of the intermediate links. It was this isolation which induced the older systematizers to institute three orders of Ungulata.

The special pedigree emanating from the Palæotheridæ includes, among the present Ungulata, the horse, tapir, and rhinoceros. The transition from the Palæotherium to the horse may be directly traced, and this, moreover, in the two most important characters, the dentition and the feet. In the Anchitherium and Hipparion, the transformation from the tridactyle to the unidactyle Ungulate is accomplished; and Rütimeyer's brilliant researches have shown how, in the milk dentition of each genus, the definitive dentition

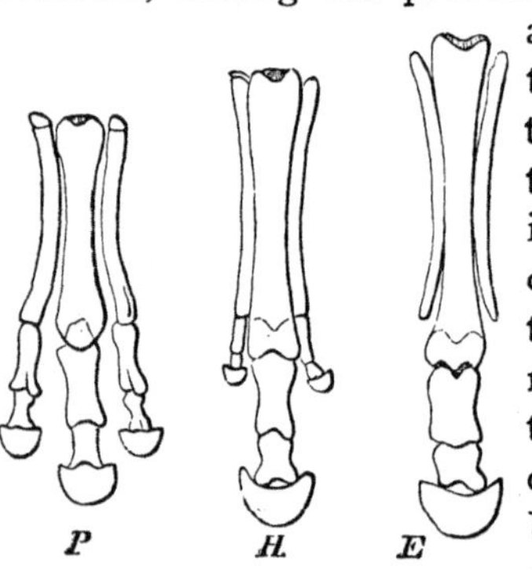

FIG. 26. Skeleton of the foot. (P) Anchitherium. (H) Hipparion. (E) Horse.

of the aboriginal genus is repeated, and Philogenesis is unequivocally expressed in Ontogenesis. The Anchitherium is a three-toed horse, in which, however, the middle toe has already undertaken the chief task. But in the Hipparion the two side toes are entirely raised from the ground, and by disuse are brought to the condition of arrest which is completed in the horse.

In the constitution of the molar teeth the tapirs have remained most faithful to the ancestral type. The circumstance that the tapir has four toes in front, whereas the Palæotheridæ known to us, have three shows, however, that the genus Palæotherium cannot have been the ancestral stock of the tapirs. For the supposition that the tapir acquired the fourth toe is contrary to all experience respecting the formation of the extremities. Rhinoceroses are also four-toed in front, and their close kindred with the tapirs is testified by the structure of their toes and a series of details in the skeleton.

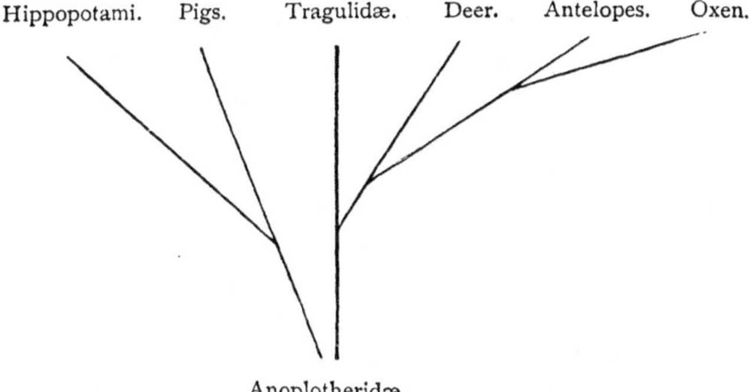

An isolated branch of the Palæotheridæ seems to be the fossil genus Macrauchenidæ, which combines the characteristics of the horse and rhinoceros with those of the camel. How far the latter, as ruminants, are directly connected with the Macrauchenidæ, or whether the form of their skull, approaching that of the horse, points to actual homology, it is for the present impossible to say.

The Anoplotheridæ are likewise distinguished by a sort of undifferentiated dentition, from which a number of specific forms might deviate in different directions. The Tragulidæ are descended from them in a direct line; they form a small group not unlike the musk animals, and are confined to South Africa and Southern Asia. As chewing the cud, they are more nearly allied to the other typical ruminants with which we are acquainted; but, on the other hand, they occupy an intermediate position towards the other non-ruminants of the division, of which the whole was united in the pre-historic world through the Anoplotheridæ. The Suidæ, or pig-like animals, were very profusely represented in the Eocene and Miocene periods. From a side branch of their predecessors, reaching up to the Anoplotheridæ, are descended the river-horses, or hippopotami. The function of ruminating is, as we know, correlated with a complex structure of the stomach as well as a peculiar mechanism of the œsophagal groove. It is naturally impossible to determine in which fossil animals these arrangements originated; yet it seems to have occurred at a very early period. Perhaps the more highly integrated structure of some non-ruminating genera, such as the hippopotamus and the peccary, may have been transmitted from the age of the Anoplotheridæ, and the very conspicuous accordance of the ruminating Tragulidæ with the Anoplotheridæ stamps the latter with tolerable certainty as ruminants.

Disregarding the camels, already mentioned as of doubtful origin, the typical ruminants separate into the deer-like and the horned. Through the hornless musk animals, the deer are connected with the Tragulidæ and the older genera. The giraffes form a side branch. But although the Helladotherium, nearly allied to the giraffe and at one time inhabiting the Athenian territory in herds, and the colossal Sivatherium, found in the spurs of the Himalayas, afford some clue to the position of the giraffes, so entirely isolated in the present world, the details of their derivation still remain very obscure.

From the antelopes to the closely allied— in fact, scarcely separable — genera of goat and sheep, and similarly to the oxen, the systematic as well as the palæontological series, and likewise the ontogenetic phases, present transitions undeniably evincing family relationship. Besides the relations of the milk dentition of the filial to the ancestral genera, which Rütimeyer has also followed minutely, great interest attaches to the gradual transformation of the skull, which reaches its extreme in the oxen, and advances from the antelope and sheep, through Ovibos, Bubalus (buffalo), Bison, to Bos (ox). In the latter, the erect position of the frontal bone attains its utmost grade, and this transformation of the skull of the antelope is repeated in the calf.

The usual classification of the Sirenia, or sea-cows, with the Cetacea, was decidedly a systematic misconception, arising from one-sided and, moreover, merely superficial consideration of the locomotive organs. All other characteristic indications—above all, the structure of the skull and the nature of the teeth—remove them from the Cetacea as much as they approximate them to the Ungulata. In the hippopotamus we have a member of this order nearly converted into an aquatic animal. We must think of the Sirenia as originally emanating from some unknown genera, which probably branched off at a very early period.

A very uncertain position is occupied by the Hyracoidæ, now represented only by a few species of the genus Hyrax. To say that their characteristics recall at once the Ungulates, the Rodents, and the Insectivora, affords no explanation. Considering the great importance of the molar teeth in deciding derivation, the chief stress should perhaps be laid on their similarity in the hyrax and the rhinoceros, and we hence regard the hyrax as an offshoot of an old Ungulate family.

With respect to the progenitors of the Proboscidæ, we refrain from any conjecture.

Later than the Graminivora,the Carnivora, and especially the beasts of prey, seem to have appeared on the scene of arctic animal life. Granting the possibility (and it is scarcely possible to do otherwise) that placental formations may have originated in various ways, the possibility likewise exists that the Carnivora, and indeed other orders too, such as the Rodents especially, may be direct descendants of carnivorous Marsupials. The oldest beasts of prey known are feline, or resemble the Viverridæ and hyenas. Then come the Canidæ, and latest of all the Ursidæ. In skull, dentition, and extremities, the seals and walruses (Pinnipedea) constitute a side branch. Although there can be no idea of any special affinity between the otter and the seal, the comparison of the two will aid us in imagining how from true beasts of prey and terrestrial animals the strange figure of seals and walruses must have proceeded.

If the conjecture already propounded should be confirmed, that the detachments and ejections of the placenta, which constitute the phenomena of the decidua, assume very heterogenous forms in groups belonging to the same family, and may be alike in others no more nearly related, the Cetacea would be installed in our pedigree in the vicinity of the beasts of prey. Between a lion and a whale an angle is enclosed, containing a countless multitude of intermediate forms. But we must always bear in mind that our business is, not to bridge over the chasms between the present peripheral ends of the series of development representing the extreme forms, but to discover the points of derivation and attachment. Fossil whale-like animals are known in the Tertiary period, such as the Zeuglodon and Squalodon. The remains of the former colossal genus are kept in good preservation at Berlin, where Johannes Müller discovered their relations to both seals and whales. The dentition is seal-like; in the skeleton there is much similarity with the whales; and although the Zeuglodons must have been preceded by a great series of species, and followed by another of considerable, if not equal, length, before the present Cetacea proceeded from them, a development of this sort seems, nevertheless, extremely probable and natural. By their still perfect dentition and the still proportionate dimensions of the skull, the Delphinoidæ are the oldest members of the true Cetacea. They were joined by the sperm whales or cachelots (Physeteridæ), and the last members are the right whales (Balænidæ). This is evinced by the fact that the whalebone or baleen plates are developed only after the rudimentary teeth have made their appearance in the jaws of the embryo, a heritage from the profusely and persistently toothed ancestors.

In the Lemuridæ, the system unites the heterogeneous remains of a collection of animals which, by reason of their prehensile hind feet with their opposable hallux, were regarded as fellow-members of the order of "true apes." The connecting link is not their anatomical constitution—they diverge widely in the form of the skull and in dentition—but rather their geographical distribution, restricted to Madagascar and a few advanced posts of Asia. Undue influence has also been allowed, certainly very unscientifically, to a certain peculiar outlandish impression which they make upon the observer. The constitution of their skull refers them to a very low grade in the scale of the mammalia. If we view them as a whole, they exhibit no general relations with any particular order of mammals, but, according to the individual genera, point to those orders which, like themselves, possess discoidal placenta; the majority of reasons favour the hypothesis that the Lemuridæ now living are the last and little modified offshoots of a division of mammals at one time far more richly developed, and that Rodents, Insectivora, Cheiroptera, and Apes, are twigs of this great branch.

The Rodents are particularly interesting, because, in conjunction with stubborn persistency in the very characteristically constituted dentition, accompanied by several peculiarities of skull, they manifest the most extraordinary power of adaptation to arboreal and steppe-life, to land and water. The Insectivora, although not nearly so rich in species, offer a similar spectacle of adaptations by which their genera have become almost repetitions of the Rodents; and the Cheiroptera (bats), in their most numerously represented division, may be regarded as a side branch of the Insectivora, if they have not proceeded directly from animals resembling the Lemuridæ.

In what geological period the monkeys were evolved from lemur-like forms we do not know. The few fossil monkeys with which we are acquainted belong to the higher families of apes, and pre-suppose a long series of ancestors. The same conjecture is forced upon us by the geographical isolation of the American monkeys from those of the Old World, which is also combined with considerable anatomical differences, although it could not occur to zoologists or comparative anatomists to deny their close systematic affinity.

The relation of the lower to the higher apes requires further discussion, which we shall combine with our disquisition on the relation of man with the monkeys.

XII.

Man.

WHEN Goethe declares, "We are eternally in contact with problems. Man is an obscure being; he knows little of the world, and of himself least of all,"[78]—he almost repeats what J. J. Rousseau says in Emile,[79] "We have no measure for this huge machine (the world); we cannot calculate its relations; we know neither its primary laws nor its final cause; we do not know ourselves; we know neither our nature nor our active principle."

Such and such-like quotations are wont to be made to us as justifying and confirming assertions of the narrowness of our powers of understanding, and of the limits of science. But in Anthropology we cannot possibly attribute any greater authority to the worthy J. J. Rousseau than to a Father of the Church; and to the Goethe, whose casual utterances are transmitted to posterity by Eckermann, we oppose the other Goethe, who in the fulness of youthful vigour, exclaims—

> Joy, supreme Creation of Nature, feeling the power
> All sublimest thoughts, which lifted her as she made thee,
> In thyself to re-echo—— *

and who conceives the most beautiful organization, as he

* Freue dich, höchstes Geschöpf der Natur, du fühlest dich fähig
Ihr den höchsten Gedanken, zu dem sie schaffend sich aufschwang,
Nachzudenken ——[80]

designates man, to be in perfect harmony with these sublimest thoughts.

Our previous reflections and deductions would lack their conclusion were man to be excluded,—could not and must not all that is said of the genesis and connection of animal being, be directly applicable to the knowledge of his nature also. The repugnance to the doctrine of Descent, the doubt with regard to it, the indignation lavished upon it, are all concentrated on its applicability and application to man; and if the body be perforce abandoned to us, the mental sphere of man is at least to remain inscrutable, a *noli tangere* to the investigation of nature. A few years ago, it was a consolation to the opponents of the doctrine of Descent that Darwin had not directly pronounced himself with respect to man. Anger was vented on his adherents, who had out-darwined Darwin. To this was added the unfortunate misapprehension that the champions of the doctrine of Descent made the human race proceed from the ennoblement of the orang, chimpanzee, or gorilla—in short, from extant apes.

But from the first appearance of the Darwinian doctrine, every moderately logical thinker must have regarded man as similarly modifiable, and as the result of the mutability of species; and Darwin has now told us, in his work on the "Descent of Man," why he did not enunciate this self-evident inference in his first book; he did not wish thereby to strengthen and provoke prejudice against his view. Knowing human weakness, he withheld the conclusion. "It seemed to me sufficient," he says, "in the first edition of my 'Origin of Species,' that by this work 'light would be thrown on the origin of man and his history,' and this implies that man must be included with other organic beings in any general conclusion respecting the manner of his appearance on this earth."

Nay, Darwin himself has now gone further, and, to the terror of all who can scarce imagine man except as created shaven and armed with a book on etiquette, he has sketched a certainly not flattering, and perhaps in many points not correct, portrait of our presumptive ancestors in the phase of dawning humanity.

Before we seriously discuss this serious subject, we will take leave to quote a more superficial verdict given by a clever essayist.[81] "Let us suppose, merely as a joke, that Nature, which we see everywhere advancing from the most simple to the complex, from the lower to the higher, had not suddenly waived this law in the presence of man; that she had not suddenly given up her evolution for his sake; that she had not suddenly begun in him a new creation; but that here, as elsewhere, she had proceeded quietly, gradually, naturally, and that man were thus nothing more than the last link of the interminable series of animals, nothing more than a 'developed ape.' The first thought that would then obtrude itself upon us, would be that the facts were not altered in the slightest degree; that man would remain as he is, with the same shape, the same face, the same gait, the same gestures, the same dispositions, powers, feelings, thoughts, and with the same dominion over the apes as heretofore. This is very simple, very self-evident, but also very important. For it confers on him—on man—the powerful sensation that, as he now is, he is a being of a quite peculiar kind, very different from even the most kindred creature; and, moreover, that this peculiar nature is his most peculiar property, whether he received it as a ready-made gift, or worked it out laboriously from a lower condition in tens of thousands of years. But if his present constitution is not in the slightest degree injured by his (assumed) animal origin, neither can his aims and tasks, his endeavours and vocations—in short, his whole future—be any other than, from his entire nature, he must imagine and believe it to be. Or must the cultivated portion of mankind be really so profoundly dismayed by the idea of descending from apes, that in despair at the impossibility of maintaining and improving the civilization, which by no means fell into their lap like ripe fruit, but which was painfully acquired, they would abandon their business and pursuits, their forms of law and government, their arts and sciences, and sink to the level of the Australian bushmen—that they would let go that by which they had raised themselves so far above the apes, and by which they are constantly raising themselves still higher, merely because it was once difficult to raise themselves above these apes even by a hair's breadth? But what man destined by nature for a ruler, would have refused to grasp the crown because his father was a hind? Or what born Raphael would have forsworn palette and pencil, because his parent had been a sign-painter? Mankind, like each individual, will use and improve its powers because it has them, not because it has obtained them from hither or thither."

We give these transient fireworks their due, but we require more profound arguments whence to derive the final verdict. To the votaries of the doctrine of Descent, its application to man is a simple deduction from a general law, gained by the method of induction. As Goethe postulated the inter-maxillary bone in man even before he had seen or proved it, so must the doctrine of Descent extend to man all its results and more or less plainly demonstrated laws. The deduction is effected by the accumulated observations of comparative anatomy, evolutionary history, and palæontology, checking and confirming one another. Thus, for all who are not satisfied with belief in miracle and subjection to the hypothesis of a revelation, nothing remains but the doctrine of Descent. To apply it to man is not more hazardous, but, on the contrary, as inherently necessary, as it is for us zoologists to make use of it in judging some polype hitherto unknown, a star-fish or a mouse. This our adversaries deny. Man, they say, has qualities which separate him absolutely from the animal, and, assuming the doctrine of Descent generally, preclude its applicability in this one case. To this assertion, so frequently to be heard, we will, in the first instance, oppose a general remark as to the apprehension of human nature.

It is commonly overlooked that, quite regardless of the validity of the doctrine of Descent or even of its existence, there is a notable inconsistency in the idea of humanity. The philosophy of history has regarded mutability, which is, in fact, capability of progress, as the essence of human nature. But if any sort of inseparable dependence of the mind upon the body be admitted, as is the case with all but an extreme spiritualistic party, the progress of mental power in mankind was inconceivable without some parallel transformation of the bodily substratum extending beyond the limits of mere variability. Even on the assumption that the mind forms its own organ, the brain, the specific idea of man would necessarily have consisted in bodily improvement, as contrasted with the supposed rigidity of the animal organism. For, in principle, it is the same whether changes take place perceptibly in arms and legs, or imperceptibly to the eye, in the molecules of the brain. We are, therefore, only retrieving the shortcomings of philosophy when we attribute to the bodily mutability of man the extension which accrues to it from the applicability of the doctrine of Descent to the particular case.

The bodily accordance betwixt man and animal leaves the doctrine of Descent so little to desire, that the apprehension of Mephistopheles lest grovelling humanity should finally be alarmed at his likeness to the Deity, might far rather be applied to his likeness to the animal. The human body, like the body of every animal, points in its evolution to an elaboration from the undifferentiated to the specialized form. The general distribution of the body and the development of the several organs is common to man and all mammals, and in the earlier stages of the embryonic state to all vertebrate animals, and indicates this general kinship. The existence therefore of a discoidal placenta (unless we prefer a special reiterated new creation of this organ of development, in which the Creator adhered to the pattern of the placenta of the lemurs, rodents, insectivora, bats, and apes) reduces us to the alternative that in the natural and to us unknown development of man, chance, or some quite different chain of causes, led in this case, as in the other. to the discoidal placenta, or that the accordance is based on consanguinity with the discoido-placental mammals. We have already (p. 272) objected to the inference that all mammalian orders are akin, should be drawn with certainty from the superficial accordance of the placenta, and we must therefore justify ourselves now, when we lay a stress on the accordance of the placenta of man and apes. The orders mentioned above all possess a placenta of small extent and discoidal form. In the shape of this disc, and in the number and distribution of the blood-vessels in the umbilical cord by which the fœtal respiration and nutrition are carried on, sundry varieties occur. Thus in the family of the Pithecoid apes, the placenta falls into two discs, whereas the umbilical cord agrees with that of man; in the American apes, on the contrary, the placenta is simple and the blood-vessels are different. In the orang and gorilla we know nothing of these organs, but the chimpanzee agrees with man, in that it has a simple discoidal placenta with two conducting (arteriæ umbilicales) and one reconducting vessel (vena umbilicalis).

With a general similarity of the human placenta with that of the discoido-placental mammals, man is specifically nearer to one at least of the so-called Anthropoid apes, than this one is to the other apes. And thus the constitution of the placenta is certainly of great importance in discriminating the systematic position of man. Enormously improbable as is the chance contemplated above, equally probable and solely credible is consanguinity; and with regard to general organization, in any specific comparison of man with the mammalia, the apes must occupy the foreground.

This comparison has been admirably conducted by Huxley and Broca.[82] The latter has set himself the task of investigating, solely as a descriptive anatomist and zoologist, regardless of all dispute as to principle, and undisturbed by the doctrine of Descent, whether the anatomical constitution of man, as compared with that of the ape, justifies, on general zoological principles, the union of the two in a single order—Primates. Huxley proves that the anthropomorphous apes (gibbon, chimpanzee, orang, gorilla) differ from the lower apes much more than from man; and that if we are obliged to assume the reciprocal consanguinity of the apes, the common derivation of the anthropomorphous apes and man is at least equally natural.

Between the peripheral members of the systematic groups of monkeys—for instance, between the American Sahuis and the Old-World Pavians and Anthropomorpha —notable differences exist in the constitution of the limbs and other parts of the skeleton, together with the soft parts belonging to them, in the muscles especially, as well as in dentition and the structure of the brain. It is false to call apes quadrumana, for within the order of the apes the contrast between hand and foot makes its appearance in its essential anatomical attributes, and in the anthropomorphous apes, in the gorilla especially, it is almost as distinct as in man.

Lucä, the anatomist renowned for his careful measurements of the cranium, imagines that he has discerned a highly important demarcation between man and the ape. In the ape, the three bones forming the axis of the skull, the basi-occipital bone, and the two sphenoid bones, lie almost in a line, whereas in man there is a double flexure of this axis; moreover, in the apes the angles increase with age, which in man decrease, and *vice versâ.* Likewise in man the occipital foramen becomes more horizontal with age, more vertical in the ape. But all this shows only, what the doctrine of Descent asserts, that the two series, ape and man, diverge from one another, and that the youthful individuals are more alike than the older ones,—that the ape as he grows becomes more bestial; man, as the riddle of the sphinx already intimated, more human. The flexure of the basal bone and the horizontal position of the occipital foramen occasions the upright gait, wherewith the differentiation between hands and feet is completed. This flexure of the cranial axis may therefore still be emphasized as a human character, in contradistinction to the apes; the peculiar characteristic of an order can scarcely be elicited from it; and especially as to the question of Descent, this circumstance seems in no way decisive.

Not only as regards hand and foot, but also in dentition and brain, the anthropomorphous apes approach man much more nearly than they do the inferior wide-nosed monkeys of the New World. These, have six molar teeth, and their brain displays the imperfections of the brain of the lemurs and rodents. Like the monkeys of the Old World, on the contrary, the anthropomorphous apes possess five molar teeth, and every portion of the human brain, even to the hippocampus minor, is likewise present. The dispute as to this insignificant portion of the brain, which R. Owen claimed as an exclusively human characteristic, possesses a merely historic interest, since, in conjunction with the posterior corner of the lateral ventricle, it has been exhibited by a number of distinguished anatomists in the orang and chimpanzee.

Thus, for those who will not relinquish their hope of finding specific distinctions between the brain of man and ape, there remain only the furrows and ridges on the surface of the cerebrum, the so-called convolutions of the brain. But here, again, it is in vain to look for fundamental differences, unless the chief stress is to be laid upon the circumstance that in the human embryo the folds commence in the frontal, in the apes in the supra-orbital lobes. The constant convolutions common to all human brains are seen in the orang and chimpanzee. These convolutions are lost, or rather exist in less perfection, in the apes approaching the Anthropomorpha; they are totally absent in the Ouistitis. But so great is the resemblance of the brain of the two apes mentioned, with that of man, that, as Broca says, "it requires the eye of an experienced anatomist to discriminate, in drawings reduced to the same dimensions, their brain from the human brain, especially if the object of comparison selected, be the brain of negroes or Hottentots, which are more simple than those of white men." A desperate attempt to rescue a specifically human cerebral character was made by the lamented Gratiolet, the anatomist, of Paris. Man was to be distinguished by the so-called transitional or bridging convolutions. These transitional folds are convolutions, by which the posterior lobes of the cerebrum are joined to the anterior and lateral portions. But Broca has lucidly demonstrated that it is the same with this as with other characteristics, and that the transitional folds in the orang, for instance, are far more like those of man than of the chimpanzee, and that the differences which exist can at the most have the value of specific or generic characters.

The distance between the lower and higher apes is far greater than between the latter and man; and if the consanguinity of the entire apedom is decisive in favour of Darwinistic views, there can be the less doubt of the kindred connection of the Old-World apes to mankind. But the form of the mature skull and of the dentition (to lay a stress upon these organs), preclude the idea that the direct ancestors of man are to be found among the apes now living. The cheap jest, produced with so much glee, of inquiring why we do not behold the interesting spectacle of the transformation of a chimpanzee into a man, or conversely, of a man by retrogression into an orang, merely testifies the crudest ignorance of the doctrine of Descent. Not one of these apes can revert to the state of his primordial ancestors, because, except by retrogression—by which a primordial condition is by no means attained—he cannot divest himself of his acquired characters fixed by heredity; nor can he exceed himself and become man; for man does not stand in the direct line of development from the ape. The development of the anthropoid apes has taken a lateral course from the nearest human progenitors, and man can as little be transformed into a gorilla as a squirrel can be changed into a rat. Man's kinship with the apes is, therefore, not impugned by the bestial strength of the teeth of a male orang or gorilla, or by the crests and protuberances on the skulls of these animals. A renowned zoologist, one of the few who adhere to the old belief, has taken the useless trouble of proving that the skull of the orang could not possibly be transformed into the human head. As if the doctrine of Descent had ever asserted such nonsense! The bony skull of these apes has reached an extreme, comparable to that of our domestic cattle. But this extreme appears only gradually in the course of growth, and the calf knows little of it, but possesses, as we have already mentioned, the cranial form of its antelope-like ancestors. In the present antelopes, and likewise in goats and sheep, this form, transitory in the calf, has remained stable. Now, as the youthful skull of the anthropomorphous apes exhibits, with undeniable distinctness, a descent from progenitors with a well-formed and still plastic cranium, and a dentition approximating to that of man, the transformation of these parts in conjunction with the brain, the latter by reason of its persistently small volume, has, as it were, struck out a disastrous path, while in the human branch, selection has effected a higher conservation of these cranial characters.

With this falls also the objection recently raised by the venerable Karl Ernest v. Baer, that it is inconceivable how, from the monkey's feet, arranged for climbing and grasping, the human foot, adapted for flat treading and walking, should be evolved in the struggle for existence. The tendency to oppose the big toe to the others, that is, to a prehensile foot, is known to be a human attribute, and this tendency is certainly inherited. How far the capacity for climbing may have been developed in the primordial ancestors, is as much unknown as these primordial ancestors themselves. Thus the aptitude in climbing shown by most of the present monkeys is only remotely connected with the inaptitude of man, and as a criterion of consanguinity, can hardly be taken into consideration.

While requiring by logical deduction, a common origin for man and the anthropomorphous apes, the doctrine of Descent, as it is almost superfluous to say, repudiates the senseless demand for intermediate forms between man and the gorilla. What future times may perhaps discover, are intermediate forms which go back to the common point of derivation of the present apes and of man. And thus, notwithstanding the very close relations already discussed, there remains the chasm which is approximately expressed by the comparative weights of the lowest human brain yet measured and the brain of the gorilla. The brain of a bushwoman, normally efficient after the manner of her tribe, amounted to 2 lbs. 4 ozs. (Cuvier's brain weighed 4 lbs. 4 ozs.), that of a gorilla may be estimated, from the capacity of the cranium, at about 1 lb. 6 ozs., which gives the approximate ratio of 3 : 2. But exalted above the animal as man may feel himself in his bodily nature, in this again he forms no exception, as many animal forms occupy an equally isolated position with reference to their unmistakably nearest kindred.

Need we imagine a twofold creation of vertebrate animals, because the lancelet is now separated from the fishes by a whole scale of intermediate forms no longer extant? The example of the horse is, among others, highly instructive in this case. Let us bear in mind that, in the nature of the limbs and teeth, this genus differs far more from all other extant graminivora than man differs from the ape. Had not the fossil ungulates been found which demonstrate the common origin of the horse with the didactyles and multidactyles, we should still not deem the horse a special miraculous creation, but incontrovertibly deduce his true kinship with the other ungulata. This pure deduction is not requisite, as the progenitors of the horse are present in conspicuous remains; and, as we have already seen, elicited in R. Owen, half a century ago, the conviction of a direct metamorphosis of the tridactyle genera into the unidactyle. Our acquaintance with the tridactyle horses was a lucky chance; they were indigenous in those parts of Europe which have been most diligently laid bare and explored in behalf of Palæontology.

But that our museums are still destitute of the fossil progenitors of man, is not more strange than the deficiency, hitherto existing, of intermediate forms, which, for example, would conclusively decide the position of the Dinotherium in the system. We will also refer again to the elephants, who, with their nearest ally, the mastodon, occupy towards the other Pachydermata a position elucidated by no fossils, and far more isolated than that of man to the apes. We hope herewith to have shown that the argument that, by peculiarities not bridged over,—by upright gait, comparative hairlessness, chin, preponderance of brain, &c.,—man betrays a position absolutely apart, cannot be admitted by comparative anatomy and palæontology. The demand, therefore, that the adherents of the doctrine of Descent should produce the intermediate forms which at one time necessarily existed, can be made only by dilettantes to whom the province of life, as a whole, has remained a sealed book.

As we observed before, the bodily nature of man is sometimes ceded to natural inquiry as a means of more certainly rescuing the other side of the dualism. But here, too, we will not be defrauded of our say and our own opinion. The mental powers of man, in their origin, growth, and effects, are likewise susceptible of investigation, and psychology only too long thought it possible to elude physiology. Let us, therefore, proceed in good heart to a short examination.

It is universally admitted that a certain relationship, or analogy, exists in the psychical capacity of the higher animals and man. Reason alone, it is said,—the essence of psychical agencies by which man attains self-consciousness, and rises to abstract conceptions, combines ideas, especially religious ideas, and lives in art and science,—this the animal does not possess. We reply that animals certainly do not possess this degree of mental development, but neither does man possess it in lower phases of evolution.

The soul of the new-born infant is, in its manifestations, in no way different from that of the young animal; these manifestations are the functions of the infantine nervous system; with this they grow and are developed together with speech. The grade to which this development rises is generally dependent on the preceding generations. The psychical capacities of each individual bear the family type, and are determined by the laws of heredity. For it is simply untrue that, independently of colour and descent, each man, under conditions otherwise alike, may attain a like pitch of mental development. As a proof of this primary equality of mankind, single instances of gifted negroes and Indians are held up to us. But these have behind them unnumbered generations practised in multifarious employments, skilled in human intercourse, even if it be one-sided ; and if these rare phenomena are thoroughly investigated, they still remain behind the average individuals of advanced races. Now it is certain that in each race, each individual passes through grades in the scale of mental development, which, in perfect analogy with the laws of anatomical development, are universally valid ; whereas, as we have seen, the psychological peculiarities of the race become valid. But it is in mankind as in the individual ; in the lapse of time, it has acquired those higher powers of mind which we call Reason.

History shows, and no one denies, a mental advance, but only in nations which have taken part in history, and only so long as this part and the exercise of the mental organs has been continued. But inferior human races exist—we may also call them human species—which are related to the others, as are lower animals to higher. It might even be given as the characteristic of the genus man, that its species occupy such extraordinarily different grades of mental condition. We are not misled by the contrary statements of missionaries and other philanthropists ; by the talk of human dignity and divine resemblance ; nor do we seek for consolation in the development still to be expected in all nations which have hitherto lagged behind. It is indeed self-evident from the theory of descent and selection, that many of the races now standing far behind in a mental point of view will in future have made a great advance. But for others, if we contemplate the ethnology and anthropology of savages, not from the standpoint of philanthropists and missionaries, but as cool and sober naturalists, destruction in the struggle for existence as a consequence of their retardation (itself regulated by the universal conditions of development), is the natural course of things.

If we examine the mental condition of mankind, and compare it with the psychical capacities of animals, we must not take the average Indian or European as our standard, but the Australian and Papuan races, which in body also have remained at a grade which the other more favoured races outgrew in pre-historic times long past. Many, it is true, overcome all difficulties, inasmuch as, assured of an equalizing human dignity as of a dogma needing no further foundation, they are ready with the assertion that it is impossible to doubt that they have retrograded from a higher mental development and sunk into barbarism. But if this possibility might be admitted as regards some few races, such as the Fuegians; in others, in the Australians for instance, any real evidence of this previous more elevated condition is wanting.

The superior mental prerogatives which are supposed to separate man from the animal, hinge more or less on the following points.

Man alone, it is said, is capable of development or progress. Specifically, human is all progress regulated and effected by human speech, for many animals likewise possess the gift of communication. But if we are not to imagine man as having advanced from all eternity, the question is, how was this advance initiated, and the whole concern is fundamentally reduced to the problem of the origin of language. We will return to this subject. Progress in general is not however to be denied to the animal. Who can question that some canine races, of which the descent from stupid jackals and wolves is as good as certain, have raised themselves mentally far above their ancestry? Who, that has read the comprehensive investigations of H. Müller, the brother of our Fritz Müller, can doubt that the honey-bee, as it gradually attained its bodily advantages and peculiarities, developed likewise the higher mental powers, corresponding with the more minute and complex organism of her brain? Man—such is the thesis we propound, reserving the question of language—differs from many animals only in the degree and means of progress. It is therefore unscientific to contrast humanity and animality in the abstract.

Man alone, it is further maintained, has a free will. In so far as the more highly developed man acts in accordance with philosophical, moral and religious principles, for which he is indebted to education and instruction—in so far as he is able to apprehend ideals, and strive after them with his own mental and bodily power, this command of will may be readily admitted, although we know that this "freedom" is likewise the collective result of natural causes. But the more simple and uniform the conditions of life, the more do the dealings of men lose the semblance and character of freedom, and the more does the individual act after the will of the tribe—I might say, of the herd—that is to say, instinctively. In this case actions are not performed even with the astounding premeditation with which some few happily organized individual animals of some few species turn the circumstances to account with apparently complete free will. The free will of the morally elevated man, is no common property of all mankind.

Man alone, and all men, are supposed to have a conscience. We consider, on the contrary, that conscience, which is known to be utterly lost in many individuals of even the most civilized nations, is, like moral will, a result of education in some few races and tribes. Fear of detection after a bad action, is not conscience; and that well-trained dogs have sensations of conscientious shame far superior to the animal terror of savage cannibals after they have wrought the murder of their fellow-men, it is impossible to deny. Of this, evidence in profusion is accumulated in the anthropological compilations of Waitz.

That a consciousness of the Divine existence is a fundamental property of all men, we likewise hold in question. It is, again, an established phrase that the most barbarous nations are guided by emotions and cravings, however obscure, towards the unknown God. This assumption is as old as the well-known attempt to prove the existence of God, "*De quo omnium natura consentit, id verum esse necesse est*" (That in which all intuitively agree, must necessarily be true). How often has this saying of Cicero been thoughtlessly repeated? This idea of God is, however, as little intuitive as the discrimination of good and evil by the conscience. Others maintain the contrary. Thus Gerland says of the Australians:[83] "The statement that Australian civilization indicates a higher grade is nowhere more clearly proved than here (in the province of religion), where everything resounds like the expiring voices of a previous and richer age; but we in no way receive the impression that we are dealing with stagnation or incomplete development. Thus the idea that the Australians have no trace of religion or mythology is thoroughly false. But this religion is certainly quite deteriorated, and has degenerated into a wild, disjointed, and often incredibly absurd demonology, into a superstitious fear of apparitions."

But when a few lines later in the work quoted, we are informed that the natives to the west of the Liverpool range, ascribe everything in nature which they cannot explain to the Devil-Devil, and that this is manifestly only a name, derived from the English Devil, for a Deity of whom they have not preserved any distinct conception, the shallowness of this evidence in favour of the hypothesis of a previous standpoint, now sunk into oblivion, enables us to infer the value of the other instances. We have far more reason to believe this low state of mental development in harmony with the bodily condition, when we hear that the natives of the Gulf of St. Vincent and the neighbourhood of Adelaide are extremely hairy, and that even the brown-coloured down of the children is so abundant and so long, that the skin of boys of five or six years of age assumes a furry appearance. But, contrary to all experience and history, we are required to believe [84] that the inhabitants of the northern parts of Australia are the most aboriginal, for "they are the most civilized, as well as the best developed, in mind and body; they only are fixed in one dwelling-place; and in any case the supposition is easier and more natural that the other natives should have degenerated, with their eternal wanderings, than that the former, fixed by the more convenient territory, should have raised themselves."

This inverts all that has hitherto been called anthropology. Moreover, there are even very advanced nations without any consciousness of God. Schweinfurth relates that the Niam-Niam, that highly interesting dwarf people of Central Africa, have no word for God, and therefore it must be supposed, not the idea; and Moritz Wagner has given a whole selection of reports on the absence of religious consciousness in inferior nations. When, in spite of all these corroborations, it is always retorted afresh that even among the lowest savages some sort of feeling of superior powers is manifested, the dispute finally results in mere verbal criticism, which has no farther interest for the doctrine of Descent.

And yet we cannot leave this subject without alluding to a fact, universally known, but, strange to say, not as yet employed in this connection, and which, as it would seem, is by itself sufficient to invalidate the assertion that the idea of God is immanent in human nature. We mean the fact that many millions in the most cultivated nations, and among them the most eminent and lucid thinkers, have not the consciousness of a personal God; those millions of whom the heroic David Strauss became the spokesman when he adopted for his own the motto of his favourite, Ulrich von Hutten: I have dared it—Jacta est alea!

And now as to Language? All modern philologists agree that languages are developed, and that most probably all linguistic families pass through three stages. In the stage of the radical languages all words are roots, and are merely placed side by side. In the second stage, that of the agglutinated languages, one root defines the other, and the defining root ultimately becomes merely a determinative element. Finally, in the inflected languages, the determinating element, of which the determinating significance has long vanished from the national consciousness, unites into a whole with the formative element. As we have said, this development, in which retrogression takes an extensive share, is universally admitted. Opinions differ only as to the origin of the linguistic material, which the acuteness of the philosophers extracts in the guise of "roots." A great authority, Max Müller,[86] discerns in the existence of the roots evidence of the absolute separation of man from the animal. While Locke says that man is distinguished from the animal by the power of forming general ideas, the philologist ought to say that human language is distinguished from the animal capacity of communication by the power of forming roots. To trace up all words to imitation and exclamatory sounds is inadmissible, as we most frequently come upon roots of fixed form and general meaning which are inexplicable in themselves. He deems the existence of these ready-made roots, before which linguistic science stands helpless, an insurmountable impediment to the apprehension of man as a link in the general evolution of organisms.

This point excepted, this excellent scholar naturally admits all those phenomena of heredity, acquisition, and degeneration, which are manifested in the laws of language, and find their most perfect analogies in our doctrine of Descent. If, for instance, we compare Zend with Sanscrit, and hear several of its words explained, we are at once reminded of the rudimentary organs and their significance. A host of anomalies are, like the isolated organisms of present times, primæval and peculiarly normal remnants and witnesses of bygone linguistic periods. In short, down to the minutest details, linguistic research stumbles on accordance and analogies with the doctrine of the derivation of organisms. And, forsooth, we are to halt before the origin of language as before a something incomprehensible and inscrutable!

This is not done, however, by the majority of comparative linguists in the present day. Though Max Müller calls the roots "phonetical fundamental types produced by a power inherent in human nature," though, according to him, man in a more perfect state possessed the power of giving to the reasonable conceptions of his mind a better and more subtle expression, the talented Lazarus Geiger[87] terms the hypothesis of a now extinct power of forming languages, and the other hypothesis connected with it, of a primordial state of higher perfection, a recourse to the incomprehensible and a return to a standpoint of mysticism. For that which is not understood is not necessarily incomprehensible. It is not our business to side with Geiger, who attributes an essential share in the ejaculation of words to the visual perceptions, or with Bleek, G. Curtius, Schleicher, Steinthal, and many others, who assign to the imitation of sounds the first place in the evocation of language. This much is, however, certain, that although those who are not critical, find Max Müller's standpoint highly convenient, in science, it is unique. In this province, interwoven as it is with the investigation of nature, the greater number of authorities, on linguistic grounds, comparative and philosophical, have been forced to the conclusion that, from an irrational primordial state, man-like beings gradually became human, while with language, the work of many years, reason made its appearance.

As early as 1851, when the doctrine of Descent was still unheard of, Steinthal[88] says: "As language arises, mind originates." Ten years after Darwin, Geiger writes: "Language created reason; before language, man was irrational." To him, and to all who have abandoned the standpoint of mysticism, "man is a genus springing from an animal condition by means of the origin and unfolding of his idiosyncrasy. And this conclusion is not, as orthodoxy and reaction are anxious to impress upon the multitude, borrowed from Darwinism, but deduced from linguistic inquiry in its own way, only by a scientific method. It need only be indicated that, as Geiger has historically proved in so many instances, "slow development, the emergence of contrast from imperceptible deviations, is the cause that the same word acquires various meanings;" that the creation of language therefore rests upon this process, and nowhere makes its appearance suddenly and abruptly; that the so-called laws of sound are habits of sound; that the special meaning which a sound has acquired in lapse of time is always the result of mere chance, or, in other words, of development.

This deduction of linguistic inquiry most fully confirms the result of natural inquiry. And any one who takes the trouble to follow the course of linguistic science will be convinced that its champions, except, perhaps, Bleek, Schleicher, and Friedrich Müller, are labouring rather to discredit, than to acknowledge, the influence of the doctrine of Descent. All the higher is our estimate of it, and therewith the most powerful objection to the inclusion of man in the great law of derivation is set aside.

The rest is incidental and a matter of detail. The question so often ventilated, and now thoroughly worn out, whether mankind is descended from one or more pairs, is solved by the inference that the stock in which language first arose, separated itself gradually from its animal progenitors, and that the selection which led to language and reason necessarily took place among large communities of individuals. The scriptural conception of the unity of the human race would be more nearly approached if all linguistic families pointed to a single source. But if it could be shown that certain linguistic families lead to utterly discordant roots, the investigation of nature might furnish the inevitable corollary that language originated in various parts of the world,—in other words, that a separation into species took place before selection had reached the point of forming language. The latter case is by far the most probable, and is, in fact, received as the only one possible by most of the linguists occupied with this question, and is most especially defended by Friedrich Müller.[89] "At the time," he says, "when there were races and no nations, man was a speechless animal, as yet, entirely destitute of the mental development which rests upon the agency of language. Independently of the premisses unfolded by natural history, this hypothesis is forced upon us by the contemplation of the languages themselves. The various families of languages, which linguistic science is able to discriminate, not only presuppose, by their diversity of form and material, several independent origins, but, within one and the same race, they point to several mutually independent points of origin."

In order to afford the reader some notion of the connection of the families of nations, we give the subjoined pedigree, in which Friedrich Müller closely adheres to Haeckel's sketch.

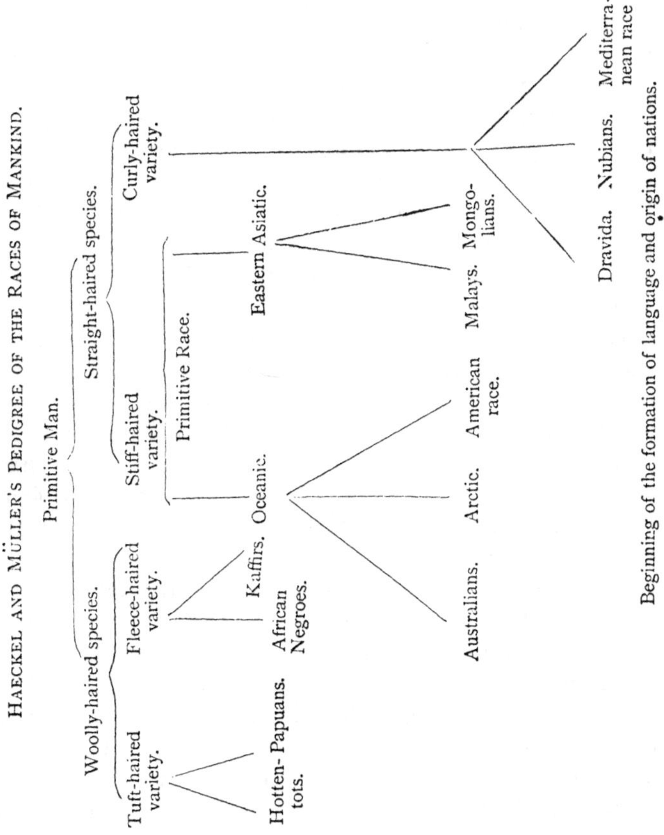

It makes mention of species and races of mankind, the species being regarded as no longer existing, while the present forms of man are distinguished only as races. On this subject, we shall not lavish many words, since, examined in the light, it is an affair of words only. In the order of Primates, man constitutes a single family, and represents it by a single genus. Whether Negroes, Caucasians, Papuans, American-Indians, &c., be called species or races, matters little. The facility of intercrossing the different nations would favour their characterization as races; but as the crossing of species does not differ in principle from the crossing of races, and as to the bodily varieties displayed in colour, hair, skull, limbs, and other characters are added the profound differences of language, the division of the genus homo into species, diverging into many races, seems after all more natural. But ultimately, as in the question of species in general, the individual feeling of each person proves decisive. Whether it was a lucky hit to found the division of mankind on the position of the hairs, in tufts or equally distributed upon the scalp, and furthermore on the section of the hair, whether it be more flat and oval or circular in form, and finally on the inclination to curl or to lie stiff and smooth, the future must decide.

The twelve races cited in the table given above, may be characterized by the aid of natural history; and as within the limits of the best known races, languages and families of languages may be found, which preclude any common origin, it follows that the formation of language began only after the still speechless primordial man had diverged into races. In geological periods and primordial history, all chronology is extremely deceptive: we may, nevertheless, acquiesce in an estimate made by Friedrich Müller as to the development of the languages of the Mediterranean races. The linguistic families of the nations dwelling chiefly in the basin of the Mediterranean are Basque, Caucasian, Hamito-Semitic and Indo-Germanic languages. "The languages of these four families," says Friedrich Müller, "are, as is generally accepted by the most competent linguists, not mutually related. If we therefore see that the Mediterranean race includes four families of people in no way related to one another, the inference is obvious that, as each language must be traceable to a society, the single race must have gradually fallen into four societies, of which each independently created its own language. A further inference is, that the race, as such, does not acquire a language; for, were this the case, race and language would now be coextensive, which is not the case.

"We must therefore assume that at the time when the various nations of the Mediterranean race were one,—the time when man belonged to no nation, but merely to a race,—mankind was destitute of language. Müller considers 3000 years approximately sufficient for the period elapsing between the divergence of the race into still speechless societies, and the epoch at which they formed nations, separated and characterized by languages; a period which might seem to many, estimated as far too short. If the ancient civilized people of Egypt be now added on, and the period of its conjectured migration from Asia computed, "the year 6,500 before the commencement of our chronology seems to be the earliest epoch at which we may speak of a Hamito-Semitic primæval people in the north of Europe." Therefore a Mediterranean race already existed 12,000 years ago. But what space of time was requisite to enable primitive man to separate into races, is entirely beyond computation, and the more so as not the slightest trace of him has hitherto been found.

With the invariable testimony of Geology that the periods of the terrestrial strata imperceptibly merged into one another, and that, especially from the Tertiary, through the Diluvial period, to the present age, continuity has been only locally interrupted, the question of the "fossil man," formerly looked upon as cardinal, has assumed another aspect. In Europe, man lived with the mammoth and the rhinoceros with a bony nasal partition (Elephas primogenius, Rhinocerus tichorhinus). It has been asserted that European man existed as early as the upper Tertiary age, but the evidence is disputable. Such remains as we have of this oldest man known to us, display a high grade of development, and certainly belong to the period at which man had already found in language the implement wherewith gradually to free himself from the dross of his lowly origin. Whether the primitive man be found or not, his origin is certain.

REFERENCES AND QUOTATIONS.

[1] Luthardt, Apologetische Vorträge. 7 Vortrag. P. 129.

[2] Philosophia quaerit, theologia invenit, religio possidet veritatem.

[3] Tageblatt der Naturforscher-Versammlung in Leipzig, 1872. P. 12. The discourse was also printed separately.

[4] A. Fick, Physiologie, 1860.

[5] Any one who wishes to be more deeply instructed in the problem of sensation, as an universal primary characteristic of the constituent elements of matter, may be referred to the very lucid and interesting work, " Das Unbewuste vom Standpunkt der Physiologie und Descendenztheorie" (Berlin, 1872). Published anonymously.

[6] L. Geiger, Ueber den Ursprung der Sprache. (Stuttgart, 1869.) P. 207.

[7] Rollet, Ueber Elementartheile und Gewebe und deren Unterscheidung. Rollet, Untersuchungen, etc. 1871.

[8] Karl Ernst v. Bär, Ueber Entwickelungsgeschichte der Thiere, Beobachtung und Reflexion, 1828.

[9] Ib. I. 223.

[10] Ib. I. 230, &c.

[11] Credner, Elemente der Geologie, 1872. P. 253.

[12] Agassiz, Essay on Classification, 1858. " It exhibits everywhere the working of the same creative Mind, through all times, and upon the whole surface of the globe."

[13] Rütimeyer, Beiträge zur Kenntniss der fossilen Pferde. Verhandlung der naturforschenden Gesellschaft in Basel, 1863. III. 642.

[14] The passages are from an occasional address—Oratio de ellure habitabili—contained in the Amœnitates academicæ. " Initio

rerum ex omni specie viventium unicum sexus par fuisse creatum, suadet ratio."

[15] Ib. Non multum a veritate me aberratum confido, si dixerim, omnem continentem terram fuisse in infantia mundi aquis submersam et vasto oceano obtectam, præter unicam in immenso hoc pelago insulam, in qua commode habitaverint animalia omnia et vegetabilia læte germinaverint."

[16] Tot numeramus species, quot ab initio creavit infinitum ens."

[17] Geoffroy St. Hilaire wrote to Cuvier: "Venez jouer parmi nous le rôle de Linné, d'un autre lêgislateur de l'histoire naturelle."

[18] Ossements fossiles.

[19] L. Agassiz, An Essay on Classification, 1859. P. 253:—

"*As representatives of Species*, individual animals bear the closest relations to one another; they exhibit definite relations also to the surrounding element, and their existence is limited within a definite period.

"*As representatives of Genera* these same individuals have a definite and specific ultimate structure, identical with that of the representatives of other species," etc. See also P. 261:—

"*Branches* or *types* are characterized by the plan of their structure;

"*Classes*, by the manner in which that plan is executed, as far as ways and means are concerned;

"*Orders*, by the degrees of complication of that structure;

"*Families*, by their form, as far as determined by structure;

"*Genera*, by the details of the execution in special parts; and,

"*Species*, by the relations of individuals to one another, and to the world in which they live, as well as by the proportions of their parts, their ornamentation," etc.

[20] Haeckel, Generelle Morphologie der Organismen (Berlin, 1866). II. 323, &c.

[21] L'espèce est—"la réunion des individus descendant l'un de l'autre et des parents communs, et de ceux qui leur ressemblent autant qu'ils se ressemblent entr'eux." Cuvier, Le Règne Animal.

[22] O. Schmidt, Die Spongien der Küste von Algier, 1868, and Versuch einer Spongienfauna des atlantischen Gebietes, 1870.

[23] Haeckel, Die Kalkschwämme. Eine Monographie in Zwei Bänden, Text und einem Atlas mit 60 Tafeln Abbildungen (Berlin, 1872).

[24] Hilgendorf, Ueber Planorbis multiformis in Steinheimer Süsswasserkalk. Monatsbericht des Berliner Akademie aus dem Jahre 1866. P. 474, &c.

[25] Waagen, Die Formenreihe des Ammonites subradiatus. Beneke's Beiträge, 1869. Vol. 2.

Zittel, Die Fauna der ältern Cephalopoden führenden Tithonbildungen. Paläontologische Mittheilungen, 1870.

Neumayr, Jurastudien. Jahrbuch der geologischen Reichsanstalt, 1871.

L. Würtenberger, Neuer Beitrag zum geologischen Beweise der Darwin' schen Theorie, 1873.

[26] Darwin, The Variation of Plants and Animals under Domestication 1868.

[27] L. Oken, Die Zeugung, 1805. Lehrbuch der Naturphilosophie, 1809–11, Pt. 3.

[28] I have borrowed the following account from my essay : "Wae Goethe ein Darwinianer?" (Was Goethe a Darwinist?) Gratz, Leuschner and Lubinsky, 1871.

Also another small work of mine: "Goethe's Verhältniss zu den organischen Naturwissenschaften" (Berlin, 1852). To the passages given in the text, which might make Goethe appear as a Darwinist, I may add the following from Eckermann's "Gespräche mit Goethe" (3 Ed. p. 191). "Thus man has in his skull two empty cavities. The question why? would not go far, whereas the question how? teaches me that these cavities are remains of the animal skull, which in those inferior organisms exist to a greater degree, and are not entirely lost even in man, notwithstanding his higher elevation."

[29] A somewhat depreciative opinion of Goethe's importance in this sphere is pronounced by V. Carus in his "Geschichte der Zoologie" (München, 1872). The reader may compare: "How little, notwithstanding his repeated study of anatomy, he had gained a true insight into the structure of animals, as determined by law, is testified by his Introduction to Comparative Anatomy. He finds no other means of harmonizing the dry details of descriptive anatomy, and the morphology which vaguely hovered before him, but by indicating the idea of a primitive type for animals, which he is, however, unable to define or to render in any way palpable by more general indications. His whole idiosyncrasy

made such a type a necessity to him, not scientifically, but æsthetically," etc. P. 590.

[30] R. Owen has declared his attitude towards the doctrine of descent in the concluding chapter of his "Manual of the Comparative Anatomy of the Vertebrata." It is published separately under the title of "Derivative Hypothesis of Life and Species," 1868.

[31] Ib. . . . . "such cause being the servant of predetermining intelligent will."

[32] "No one can enter the saddling-ground at Epsom before the start for the Derby, without feeling that the glossy-coated, proudly-stepping creatures led out before him are the most perfect and beautiful of quadrupeds. As such, I believe the horse to have been predestined and prepared for man." Ib. P. 11.

[33] "I deem an innate tendency to deviate from parental type, operating through periods of adequate duration, to be the most probable nature or way of operation of the secondary law whereby species have been derived one from the other." Ib. P. 22.

[34] Lamarck, Philosophie Zoologique (Paris, 1809). In the text allusion is made to the following passages :—

"Ainsi l'on peut assurer que, parmi ses productions, la nature n'a réellement formé ni classes, ni ordres, ni familles, ni espèces constantes, mais seulement des individus qui se succèdent les uns aux autres, et qui ressemblent à ceux qui les ont produits. Or ces individus appartiennent à des races infiniment diversifiées, qui se nuancent sous toutes les formes et dans tous les degrés d'organisation, et qui chacune se conservent sans mutation tant qu'aucune cause de changement n'agit sur elles." I. 22.

"La supposition presque généralment admise, que les corps vivans constituent des espèces constamment distinctes par des caractères invariables, et que l'existence de ces espèces est aussi ancienne que celle de la nature même, fut établie dans un temps où l'on n'avait pas suffisament observé, et où les sciences naturelles étaient à peu près nulles. Elle est tous les jours démentie aux yeux de ceux qui ont beaucoup vu et qui ont longtemps suivi la nature." I. 54.

"Les espêces n'ont réellement qu'une constance relative à la durée des circonstances dans lesquelles se sont trouvés les individus qui les représentent." I. 55.

" —les considérations, et nous font voir—

" 1. Que tous les corps organisés de notre globe sont de véritables productions de la nature, qu'elle a successivement executées à la suite de beaucoup de temps.

" 2. Que dans sa marche la nature a commencé et recommence encore tous les jours, par former les corps organisés les plus simples, et qu'elle ne forme directement que ceux-là, c'est à dire que ses premières ébauches de l'organisation, qu'on a désignées par l'expression de générations spontanées.

" 3. Que les premières ébauches de l'animal et du vegétal étant formées dans les lieux et les circonstances convenables, les facultés d'une vie commençante et d'un mouvement organique établi ont nécessairement developpé peu à peu les organes, et qu'avec le temps elles les ont diversifiés ainsi que les parties.

" 4. Que la faculté d'accroissement dans chaque portion du corps organisé était inhérente aux premiers effets de la vie ; elle a donné lieu aux différens modes de la multiplication et de régénérations des individus ; et que par là, les progrès acquis dans la composition de l'organisation et dans la forme et la diversité des parties ont été conservés.

" 5. Qu'a l'aide d'un temps suffisant, des circonstances qui ont été nécessairement favorables, des changemens que tous les points de la surface du globe ont successivement subis dans leur état, en un mot, du pouvoir qu'ont les nouvelles situations et les nouvelles habitudes pour modifier les organes des corps doués de la vie, tous ceux qui existent maintenant ont été insensiblement formés tels que nous les voyons.

" 6. Enfin, que d'après un ordre semblable de choses, les corps vivants ayant éprouvé chacun des changemens plus ou moins grands dans l'état de leur organisation et de leurs parties, ce qu'on nomme espèce parmi eux a été insensiblement et successivement ainsi formé, n'a qu'une constance relative dans son état, et ne peut ainsi être aussi ancien que la nature." I. 65, &c.

" La progression dans la composition de l'organisation subit, çà et là, dans la série générale des animaux, des anomalies opérées par l'influence des circonstances d'habitation et par celle des habitudes contractées." I. 135.

" Dans tout animal qui n'a point dépassé le terme de ses developpemens, l'emploi plus fréquent et routiné d'un organe quelconque fortifie peu à peu cet organe, le developpe, l'agrandit, et lui donne une puissance proportionnée à la durée de cet emploi; tandis quel e défaut constant d'usage de tel organe l'affaiblit insensiblement, le détériore, diminue progressivement ses facultés, et finit par le faire disparaître.

"Tout ce que la nature a fait acquerir ou perdre aux individus par l'influence des circonstances où leur race se trouve depuis longtemps exposée, et par conséquent, par l'influence de l'emploi prédominant de tel organe ou par celle d'un défaut constant d'usage de telle partie, elle le conserve par génération aux nouveaux individus qui en proviennent." I. 235.

"La volonté dépendant toujours d'un jugement quelconque n'est jamais véritablement libre; car le jugement qui y donne lieu est, comme le quotient d'une opération arithmétique, un résultat nécessaire de l'ensemble des éléments qui l'ont formé." I. 342.

"Les animaux contractent, pour satisfaire à ces besoins, diverses sortes d'habitudes, qui se transforment en eux en autant de penchans, auxquels ils ne peuvent resister et qu'ils ne peuvent changer eux mêmes. De là l'origine de leurs actions habituelles et de leurs inclinations particulières, auxquelles on a donné le nom d'instinct. Ce penchant des animaux à la conservation des habitudes et au renouvellement des actions qui en proviennent, étant une fois acquis, se propage ensuite dans les individus, par la voie de la reproduction ou de la génération, qui conserve l'organisation et la disposition des parties dans leur état obtenu, en sorte que ce même penchant existe déjà dans les nouveaux individus, avant même qu'ils l'aient exercé." I. 325.

[35] The acute author of the book, "Das Unbewusste" defines instinct in essentially the same manner as Lamarck. "In this sense it may be said that every instinct is in the last instance by its origin an acquired habit, and the proverb that 'habit is second nature' thus receives the unexpected supplement that habit is also the beginning and origin of the first nature, *i.e.*, of instinct. For it is always habit, *i.e.*, the frequent repetition of the same function, which so firmly impresses the mode of action, however acquired, upon the central organs of the nervous system that the predisposition thus originated becomes transmissible." p. 182.

[36] The highly important doctrine which Lyell has substantiated

with his rich experience is also distinctly and concisely enunciated by Lamarck in the Philosophie Zoologique:—

"Si l'on considère, d'une part, que dans tout ce que la nature opère elle ne fait rien brusquement, et que partout elle agit avec lenteur et par dégrés successifs, et de l'autre part que les causes particulières ou locales des désordres, des bouleversemens, des déplacemens, etc., peuvent rendre raison de tout ce que l'on observe à la surface de notre globe, et sont néanmoins assujetties à ses lois et à sa marche générale, on reconnaîtra qu'il n'est nullement nécessaire de supposer qu'une catastrophe universelle est venue culbuter et détruire une grande partie des opérations mêmes de la nature." I. 80.

[37] Principles of Geology.

[38] In 1870, as well as in 1872, the majority in the French Academy bore this testimony to Darwin. The reiterated proposal of electing him a member was certainly not rejected until such men as Milne-Edwards and Quatrefages had made the standpoint clear to the scientific judges.

[39] Origin of Species. Fifth Ed. 1872.

The other works cited are, "The Variation of Animals and Plants under Domestication," 1868; "The Descent of Man and Sexual Selection," 2nd ed., 1871; "Expression of the Emotions in Man and Animals," 1872.

[40] Malthus (1798) investigated the conditions of the increase and decrease of human population. He finds that the rise in population is necessarily limited by the means of subsistence, and that the growth increases in proportion to the means of subsistence, setting aside some special impediments easily discovered. These impediments, which always keep the population below the amount warranted by the means of subsistence, are moral restraint, crime, and misfortune. Malthus depicts the struggle for existence without pronouncing the word; he demonstrates that the dreams of a future blissful equality of all mankind on the earth transformed into a vast garden, are based upon delusions. Each individual must much rather labour indefatigably to ameliorate his position. By the experience of breeders and gardeners he knows that animals and plants may be improved and ennobled. No organic ennoblement of the human race as a whole is perceptible, nor can the human race be ennobled save by condemning the less perfect individuals to celibacy.

These, and similar thoughts in the work of Malthus, first suggested Darwin's theory, as he has informed us.

[41] Variation of Animals and Plants. I. 100.

[42] Two treatises by A. Kerner are also very instructive with regard to the question of species: "Gute and Schlechte Arten." (Innsbruck, 1866.) And "Die Abhängigkeit der Pflanzenwelt von Klima und Boden. Ein Beitrag zur Lehre von der Enstehung und Verbreitung der Arten, gestützt auf die Verwandtschaftsverhältnisse, geographische Verbreitung und Geschichte der Cytisusarten aus dem Stamme Tubocytisus D.C." 1869. Kerner's latest work, "Die Schutzmittel des Pollens" (Innsbruck, 1873) is likewise an admirable investigation of the variability, adaptation, and formation of species.

[43] Origin of Species. 13th ed. p. 84.

[44] Origin of Species. 13th ed. p. 96.

[45] Origin of Species.

[46] P. 7. The following pages contain an epitome of the objections offered to the inadequacy of the theory of selection.

[47] Moritz Wagner, Die Darwin'sche Theorie und das Migrationsgesetz der Organismen, 1848.

[49] Nageli, Enstehung und Begriff der naturhistorischen Art. (Sitzungsberichte der bairischen Akademie der Wissenschaften), 1865. Nageli's later investigations (Sitzungsberichte der mathematisch-physikalischen Klasse der Münchner Akademie, 1872, p. 305) confirm the doctrine of descent. He shows that the gregariousness of merely allied species and their varieties proves more favourable to the formation of species than isolation. "The associated forms—of certain Alpine plants—have, as it were, reciprocally modified one another; they exhibit, to express myself thus, *a specific social type, which is different in each assemblage, and therefore in every neighbourhood.* This fact incontrovertibly shows *that the forms have altered since they were associated.*

"The specific social type consists in their showing a notable accordance in certain characteristics, while in others they represent extremes, and in these sometimes exceed all their congeners in other districts.

"From these facts it follows undoubtedly that the movement in the cenobitic forms (*i.e.* living together) is divergent. For extreme characteristics are developed in them, whereas the eremitical forms exhibit a medium in their characteristics.

"Nageli proves that since the glacial period an alteration has taken place in Alpine plants, and the manner in which it occurred."

[50] J. Broca, L'Ordre des Primates. Parallèle anatomique de l'homme and des singes, 1870.

[51] Descent of Man, p. 367.

[52] At the time at which we write, we have before us, unfortunately, only the incomplete reports of the daily papers, and the syllabus of Professor Max Müller's "Three Lectures on Mr. Darwin's Philosophy of Language."

[53] Zöllner, "Ueber die Natur der Kometen" (1 ed. p. 305).

[54] For the further instruction of the reader, we will allow another Philosopher and Naturalist to speak respecting the primordial commencement of life, to our apprehension so simply accountable. The hypothesis of origin is under discussion. In the critical examination of the "Philosophie des Unbewussten" (7) it runs thus, p. 22. The "Philosophie des Unbewussten" says, p. 558: "It is probable that before the origin of the first organisms, organic combinations existed which (p. 556) were under the influence of a damp atmosphere, abounding in carbonic acid, and of a higher temperature, light, and stronger electric influences. If these presuppositions are adopted, and the consideration added that if conditions thus favourable to primordial generation once existed, which they must have done—they probably endured during considerable geological periods—the inference is in truth inevitable that in lapse of time and with change of circumstances, these organic substances aggregated into innumerable combinations. Among these innumerable modes of arrangement, groupings and combinations, by far the greater portion must remain at the grade of inorganic *form*, because it has not attained the needful chemical composition and physical properties; a very much smaller portion of the results produced by these combinations of organic materials might perhaps transitorily approach the organic form or even actually assume it, yet without possessing the constitution necessary to maintain it permanently; a third and yet smaller portion might perhaps maintain this form for itself in the exchange of material, about as long as the approximate duration of life of one of the most primitive of the present Protists, yet lacked those properties which preserve the species by division and reproduction after the

natural extinction of the individual; a fourth portion might possess the properties requisite for self-preservation as well as for the preservation of the genus, yet lacked that peculiar tendency to vary (Philosophie des Unbewussten, p. 591), or at least that tendency to vary in the particular direction which was alone capable of leading to development into higher forms; and finally a fifth portion possessed this property in addition to the others. It is the progeny of the fourth and fifth classes of our division which still populates the ocean and the earth.* From which species of Monera proceeded the advanced development of the Infusoria; whether from one still living or from an extinct species we do not know as yet; but this much we may accept as certain, that the majority of the Protists that we still know, belong to that fourth class which is incapable of development. The persistence of the ephemeral creations of our second and third classes would naturally be secured only so long as circumstances continued favourable to their renewed primordial generation, but from the teleological standpoint the first class must be described as that of the completely abortive attempts at creation."

These, and similar more or less interesting fancies to which we attribute no great importance, are all derived from Haeckel's hypothesis of Autogony ("Generelle Morphologie der Organismen," 179 seq.), which he set up after his beautiful discoveries on the simplest organisms now existing—the Monera and the Protists. From this work we select the following passage:—"Doubtless we must imagine the act of autogony, the first spontaneous origin of the simplest organisms, to be quite similar to the act of crystallization. In a fluid, holding in solution the chemical elements composing the organism, in consequence of certain movements of the various elements among themselves, certain points of attraction are formed, at which the atoms of the organogenetic elements (carbon, oxygen, hydrogen, nitrogen) enter into such close contact with one another that they unite in the formation of a complex ternary or quaternary molecule. This primary group of atoms—perhaps a molecule of albumen—now acts like the analogous crystal-

* It is a simpler and more probable explanation that these low organisms continue to exist because there is room for them. They remain in spite of differentiation and in consequence of differentiation.

line molecule, attracting the homogeneous atoms dissolved in the mother wather; and they now likewise coalesce in the formation of similar molecules. The albuminous granule thus grows and transforms itself into a homogeneous organic individual, a structureless moner or mass of plasma, like a Protamæba, &c. Owing to the easy divisibility of its substance, this moner constantly tends towards the dissolution of its recently consolidated individuality, but when the constantly preponderating absorption of new substance outweighs the tendency to disintegration, it is able to preserve life by the exchange of material. The homogeneous organic individual, or moner, grows by means of imbibition (nutrition) only until the attractive power of the centre no longer suffices to hold the whole mass together. In consequence of the preponderating divergent movements of the molecules in different directions, two or more centres of attraction are now formed in the homogeneous plasma, which henceforth act attractively on the individual substance of the simple mould, and thereby induce its fission, or partition, into two or more portions (reproduction). Each part forthwith rounds itself again into an albuminous individual, or mass of plasma, and the eternal process begins again, of attraction and disruption of the molecules, producing the phenomena of exchange of substance, or nutrition, and reproduction."

Relying on the known peculiarities of the combinations of carbon, Haeckel has attributed to this substance the most important part in his representation of the first development of life and the physiological phenomena of the lowest organisms. This is the "carbon theory" so strongly deprecated by his antagonists. Minds would be less heated on the subject were it remembered that a refutation of this "adventurous attempt," as Haeckel terms it, to assist the idea of genesis, would not change a hair in the compulsory logical necessity of acknowledging the evocation of life by natural means. The arguments against the carbon theory have been developed, among others, by Preyer, " Ueber die Erforschung des Lebens (Jena, 1873). It is shown that carbon, in its present terrestrial conditions, points almost exclusively to organic origin, and, as yet, no source of carbon has been demonstrated adequate for the first formation of living bodies on the earth.

[35] A. R. Wallace, The Malay Archipelago (3rd ed.: London,

1872), and Contributions to the Theory of Natural Selection (2nd ed.: 1871).

[56] "The hypothesis of Pangenesis, as applied to the several great classes of facts just discussed, no doubt is extremely complex, but so assuredly are the facts. The assumptions, however, on which the hypothesis rests cannot be considered as complex in any extreme degree; namely, that all organic units, besides having the power, as is generally admitted, of growing by self-division, throw off free and minute atoms of their contents, that is, gemmules. These multiply, and aggregate themselves into buds and the sexual elements; their development depends on their union with other nascent cells, or units, and they are capable of transmission in a dormant state to successive generations.

"In a highly organised and complex animal, the gemmules thrown off from each different cell, or unit, throughout the body must be inconceivably numerous and minute. Each unit of each part, as it changes during development—and we know that some insects undergo, at least, twenty metamorphoses—must throw off its gemmules. All organic beings, moreover, include many dormant gemmules derived from their grand-parents and more remote progenitors. These almost infinitely numerous and minute gemmules must be included in each bud, ovule, spermaozoon, and pollen grain. Such an admission will be declared impossible, but, as previously remarked, number and size are only relative difficulties, and the eggs or seeds produced by certain animals or plants are so numerous that they cannot be grasped by the intellect." Darwin, Variations of Animals and Plants, II. 526.

[57] A. Rollet, Ueber die Erscheinungsformen des Lebens und den beharrlichen Zeugen ihres Zusammenhanges. Almanach der kais. Akademie der Wissenschaften (Wien, 1872).

[58] Darwin, Variations of Animals and Plants, I. 200.

[59] V. Graber, Ueber den Tonapparat der Locustiden, ein Beitrag zum Darwinismus. Zeitschrift für wissenschaftliche Zoologie. Vol. 22.

[60] Hermann v. Nathusius, Vorstudien für Geschichte und Zucht der Hausthiere zunächst am Schweineschädel, 1864.

[61] Ib., p. 108.

[62] Descent of Man, I. 412.

[63] Origin of Species. 13th ed., p. 171.

[64] Lamarck also constructed a pedigree at the end of his "Philosophie Zoologique," in which he disposes of the greater number of classes, while he attributes to the remainder another point of derivation. He thus assumes in the animal kingdom two primordial forms derived from primordial generation. His scheme is as follows :—

TABLEAU

Servant à montrer l'origine des differents animaux.

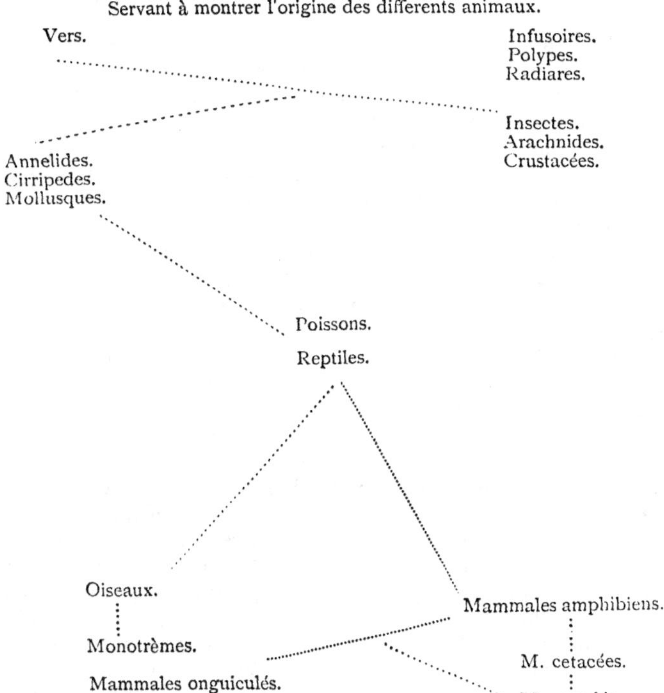

A comparison of this pedigree with the one which we now set up is extremely interesting, and shows the progress of our knowledge.

[65] Zum Streit über den Darwinismus, "Augsburger Allgemeine Zeitung," 1873, No. 130.

[66] A short preliminary communication in the "Revue Scientifique" (Paris, 1873). No. 37.

[67] Braun, Ueber die Bedeutung der Entwickelung in der Naturgeschichte (Berlin, 1872).

"The vegetal kingdom shows us—

"I. Plants, which in their vegetative development of the germ, exhibit a sexual generation, mostly in a thallus-like form. (Thallogens, Bryophytes, the Thallophytes of the authors, and Charas and Mosses.)

"II. Plants in which the first generation is transitory, and only the second develops into the vegetative, leaf-forming stem, without, however, advancing to the stage of phenogams. (Acrogens Cormophytes, the ferns, &c.)

"III. Plants in which metamorphosis advances as far as the formation of a blossom, yet without reaching the final formation, that of the formation of the carpel. (Phenogams without real fruit, gymnospermic Anthophytes.)

"IV. Plants which reach the final and highest conclusion of vegetable development, that of true fructification. (Angiospermic Anthophytes; Monocotyledons and Dicotyledons as secondary gradations.)

[68] As we have discussed in this chapter individual development with reference to historical development, we must also notice the strange opposition to the doctrine of descent offered by Kölliker. He has laid down his views in his "Monographie der Pennatuliden," and, in a separate pamphlet, bearing the title of "Morphologie und Entwickelungsgeschichte des Pennatulidenstammes, nebst allgemeine Betrachtungen zur Descendenzlehre" (Frankfurt, 1872).. Whereas Darwinism derives the continuity and harmony of the organic world from variability, natural selection, heredity, and adaptation—in short, from palpable, visibly efficacious causes—Kölliker is of opinion "that the same general formative laws which govern inorganic nature hold good also in the organic kingdom, and hence a common pedigree and a slow transformation of one form into another are entirely unnecessary for the explanation and comprehension of the accordance of the forms and series

of forms of the animate world" (p. 3). Except decided dualists, no one disputes the first part of Kölliker's thesis. But the identification of the development of the organic individual, excluding the law of heredity, with the simple process of crystallization, or any other operation of chemical combination repeating itself under given conditions, scarcely needs a detailed refutation. Kölliker says, and tries to prove, that the so-called monophyletic hypothesis, according to which the different families of organisms are derived from a single primordial form, has to struggle with insurmountable difficulties; that the hypothesis of descent from many families (polyphyletic) possesses more probability. If this be admitted, then—and here comes a bold leap of the imagination—the adherent of the polyphyletic hypothesis finds himself in a position to attribute different pedigrees and primordial forms not only to the higher divisions, but even to their genera, and to assume their independent origin. Nay, it even seems credible that the self-same species may appear in different pedigrees; as by the incontrovertible supposition of general laws of formation, it cannot be seen why like primary shapes should not, under certain circumstances, be able to lead to like final forms (see p. 21). Nay, this hypothesis does more, for "even if individuals of the same species occupy remote localities, as, for instance, Pennatula phosphorea, Funiculina quadrangularis, Renilla reniformis, &c., it is surely more fitting to assume their independent origin." Kölliker's polyphyletic hypothesis put an end to all difficulties, and, among others, it explains the so-called "representative forms" to be mentioned in our tenth chapter; for, from "this standpoint, it is credible that these forms are not genetically connected, but belong to different pedigrees" (p. 23). And all this, and much more, is supposed to be conceivable, because the world of organisms, in its consecutive development, follows intrinsic causes or definite laws of formation, "laws which, in a perfectly definite manner, urge on the organisms to constantly higher development." At the same time, Kölliker deliberates (p. 38) whether, just as here germs and buds, so also free existing youthful forms of animals did not possess the power of striking out a development different from the typical one, which freedom must be severely mulcted by the law of development, which can and must create individuals of the same species at the opposite poles. Kölliker (p. 44) thus sums up his fundamental view—"that in and with the first origin of organic matter and of organisms, the whole plan of development, the collective series of possibilities, were also potentially given, but that various external impulses operated determinatively on individual developments, and impressed a definite stamp upon them." Notwithstanding the scientific dress, dualism is here complete; whereas, Physics and Chemistry make their laws, applying to inorganic as well as to organic nature, comprehensible in their form, purport, and effects, Kölliker knows nothing of the constitution of his laws. The doctrine of natural selection allows us to recognize the causes and effects of heredity and adaptation, and establishes the phenomenal series under the form of laws. But laws which are founded only on a plan which is to be carried out prospectively and in subservience to this dower of imperfect organisms, are ignored by natural science.

[69] Ueber die Herkunft unserer Thierwelt. Einezoo-geographische Skizze von L. Rütimeyer (Basel, 1867). We have made copious use in our text of this extremely instructive writing.

[70] A. R. Wallace, Malay Archipelago. P. 10, &c.

[71] G. Koch, Die indo-australische Lepidopteren-Fauna in ihren Zusamnenhang mit den drei Hauptfaunen der Erde. (1 Ed. Berlin, 1873.)

[72] Peschl, Neue Probleme der vergleichende Erdkunde, 1870.

[73] All the more distinct is the affinity of the Mastodon and the Elephant. Between the pliocene Mastodon Borsoni and the Elephas primigenius, twenty species are interposed, among which are our still living species, the Indian and African elephants. The limits of the two genera are hereby entirely obliterated. According to other statements, the Elephas primigenius (the mammoth) falls into at least four geographical varieties, which join on to the American species. A dwarf species of elephant is found in the caves of Malta, which in dentition attaches itself to the African species.

[74] Joh. Schmidt, The Relationships of the Indo-Germanic Languages. 1872.

[75] Various antagonists of the doctrine of descent have vented their moral dismay in the most poignant expressions, precluding

any scientific discussion, on finding that the pedigree of the Vertebrata, and therewith of man, is actually traced beyond the vertebrated animals to so low a being as the Ascidians. It is otherwise with the critics of Kowalewsky's and Kupffer's observations, who acknowledge the facts, but think themselves obliged to differ in their interpretation. One of these is A. Giard, in his work on the "Embryogènie des Ascidiens." (Archive de Zoologie expérimentale, Paris, 1872.) The pupil of Lacaze Duthiers says :—"La chorde et l'appendice caudale sont chez la larve Ascidienne des organes de locomotion d'un importance assez secondaire malgré leur généralité, *pour qu'on les voie disparaître presque entièrement dans le genre Molgula*, où ils sont devenus inutiles par suite des mœurs de l'animal adulte ; l'homologie entre cette chorde dorsale et celle des vertébrés n'est donc qu'une *homologie d'adaptation* déterminée à remplir l'iodentité des fonctions, et n'indique pas de rapports de parente immediate entre les vertébrés et les Ascidiens." The author thus denies the consanguinity of the vertebrate animals and Ascidians, and traces back to adaptation the resemblance approaching identity occurring in the organs of the two. The inferences in these few sentences appear to us utterly at fault. The circumstance that in Molgula, and many other Testacea, development takes a narrower course, makes as little alteration in the importance of the facts as, for instance, the Nauplius development of the Peneus observed by Fritz Müller, or the Navicula of the Molluscs, is prejudiced by the fact that the other Decapods have forfeited the Nauplius phase, or the Landsnails the navicula phase. But it is simply incomprehensible in what the identity of functions is to consist which in the Vertebrata was capable of producing the notochord, with, it is particularly to be remarked, the spinal cord (which M. Giard entirely forgets) ; and, in the other case, the "homologie d'adaptation." We, on the contrary, see these organs performing different functions, because in the one they remain of fundamental importance through life, and not in the other. Thus we conversely lay the stress on the morphological identity accompanying functional difference. M. Giard adduces no facts.

[76] T. H. Huxley, Manual of the Anatomy of the Vertebrated Animals. German Ed.

[77] March, American Journal of Sciences and Arts, February, 1873.

[78] Eckermann, Gespräche mit Goethe. II. 152.

[79] Rousseau, Emile (Œuvres, Paris, 1820, IX. 17). "Nous n'avons point la mesure de cette machine immense; nous n'en pouvons calculer les rapports; nous n'en connaissons ni les premières lois, ni la cause finale; nous nous ignorons nous-mêmes; nous ne connaissons ni notre nature, ni notre principe actif."

[80] Metamorphose der Thiere.

[81] R. Valdek in the "Presse," 1865, No. 327.

[82] Huxley, Man's Place in Nature, 1863. Manual of the Anatomy of the Vertebrated Animals.

[83] Broca, L'Ordre des Primates. Parallèle anatomique des l'Homme et des Singes. (Paris, 1870.)

[84] Waitz, Anthropologie der Naturvölker, 6 thl., p. 796. Bearbeitet von Gerland.

[85] Do. p. 708.

[86] "Augsburger Allgemeine Zeitung," 1873, Nos. 92–94. Beilage.

[87] I heard the lectures delivered at Strasburg by this scholar, "Ueber die Resultate der Sprachswissenschaft," with great interest and advantage.

[88] L. Geiger, Der Ursprung der Sprache, 1869, p. 37.

[89] Steinthal, Der Ursprung der Sprache, 1851.

[90] Fr. Müller, Allgemeine Ethnographie. Wen., 1873.

INDEX.

Agassiz, L. 11, 80, 86, 223.
Alca impennis, 186.
Amblystoma, 210.
Ammonites, 71, 74, 97.
Amniotes, 250.
Amphibians, 209, 258, 264.
Amphioxus. *See* Lancelet.
Anchitherium, 242, 273.
Ancyloceras, 216.
Anguis, 185.
Annelids, 32, 198.
Annulosa, 260.
Anoplotheridæ, 274.
Anteater, 246.
Antelopes, 273, 277.
Apes, 280, 288.
Arachnida, 33.
Archæopteryx, 72, 265.
Archæosauros, 83.
Archegosauros, 258.
Armadillos, 228, 246.
Articulata, 32, 53, 72, 201, 219.
Ascarides, 206.
Ascidians, 36, 219, 252.
Axolotl, 209.

Baer, C. E. von, 48, 191, 197, 201, 219, 293.
Balenidæ, 280.
Barraude, 11.
Bathybius, 26,
Batrachians, 72, 258, 264.
Bats, 228, 246 269.
Bavey, 212.
Bears, 243.
Beaumont, Elie de, 130.
Beavers, 247.
Bees 28, 47.
Beetles, 185.
Belemnites, 74.
Beneden, Van, 209.
Birds, 227 232, 250, 265.
Bison, 277.
Blastoids, 77.
Bleek, 304.
Bos, 277.
Bourguignat, 224.
Brachiopoda, 70.
Braun, 193, 220.
Brehm, 182, 267.
Broca, Prof. 160, 187.
Brücke, 26.
Bruta, 237, 271.
Bubalos, 277.
Buffaloes, 244, 247.
Buffon, 5.
Butterflies, 227.

Cachelots, 280.
Camels, 223.

Camper, P. 101, 119.
Campsognathus, 264.
Canidæ, 278.
Carnivora, 241, 273, 278.
Carpenter, Dr. 64, 93.
Cassowaries, 237.
Cassidulidæ, 77.
Cats, 100, 170, 234.
Cecidomyia, 47.
Cephalopoda, 70, 154, 201.
Ceratoda, 258.
Cercaria, 46.
Ceroxylus laceratus, 181.
Cestoda, 205.
Cetacea, 238, 273, 279.
Chalina, 154.
Chalinula, 154.
Cheiroptera, 280.
Chelonia, 263.
Chimpanzee, 288.
Chirotes, 185.
Cidaridæ, 77.
Cirripedes, 207.
Civets, 234.
Cladonema Radiatum, 42, 202.
Clymenia, 71.
Clypeastræ, 80.
Coccoliths, 26.
Cockleshell, 199.
Cœcilia, 259.
Cœlenterata, 31, 218.
Cœnopithicus, 245.
Coleoptera, 184.
Comatula, 80, 197.
Conchifera, 71.
Copepoda, 207.
Corals, 42.
Crabs, 150, 153, 207, 210, 233.
Credner, 64.
Crinoid, 56, 76.
Crocodiles, 75, 260, 262.
Crossopterygii, 258.
Ctenomys, 184.
Curtius, G. 304.
Cuscus, 234.
Cuttlefish, 11.
Cuvier, 30, 85.
Cyclostomi, 256.
Cytisus, 146.

Darwin, 131, 184, 248.
Deciduata, 272.
Deer, 234, 246, 275.
Delphinoidæ, 279.
Dentalium Teredo, 200.
Desor, 76.
Dicynodonta, 260.
Didelphidæ, 245.
Dinosauria, 261.
Dinotherium, 295.
Dipnoi, 37, 256, 258.
Distoma, 46, 206.
Dodo, 186.
Dogs, 99, 138.
Dubois-Reymond, 15, 20.
Duck-mole, 237.
Dujardin, 42.
Dumeril, A. 209.

Earl, G. Windsor, 230.
Echidna, 270.
Echinæ, 76, 80.
Echinoconidæ, 77.
Echinodermata, 31, 76.
Edentata, 79, 237, 246, 269, 271.
Elasmobranchii, 256.
Elephants, 80, 231, 242, 247, 269, 310.
Enaliosaurians, 74, 238, 260.
Endocyclica, 77.
Eozoon, 68.
Ephippigera vitium, 172.
Eucladia Johnsoni, 76.

Feather stars, 76.
Fick, A. 18.
Fisk, 260, 264.

Foraminifera, 93.
Forster, G. 101.
Fowls, 135.
Frogs, 258.
Fürbringer, 185.

Ganoids, 71, 256.
Gargol, 44.
Gasteropoda, 70.
Gastrula, 51, 218, 257.
Gegenbauer, 74, 256.
Gerland, 300.
Gibbon, 289.
Giraffes, 277.
Goats, 277.
Goethe, 100.
Gorilla, 288.
Graber, Von, 171.
Graminivora, 240.
Graptolites, 69.
Gratiolet, 291.
Guinea-pig, 100.

Hæckel, 40, 89, 178, 198, 218, 250.
Heer, 238.
Helladotherium, 277.
Helliconidæ, 179.
Herder, 4.
Hilgendorf, 96.
Hipparion, 273.
Hippopotamus, 242, 275.
Holothuria, 78, 197, 217.
Honeysuckers, 233.
Horses, 81, 170, 225, 242, 273, 295.
Humboldt, W. von, 4, 222.
Huxley, 257, 264, 289, 307.
Hydractinea carnea, 203.
Hydra tuba, 178.
Hydrophilus piceus, 53.
Hyænas, 243, 278.
Hylodon Martinicensis, 212.
Hypsilophodæ, 265.
Hyrax, 269, 277.

Ichthyornis dispar, 267.
Ichthyosauria, 74, 260.
Inflata, 216.
Infusoria, 269, 280.
Insecta, 32.
Insectivora, 269, 280.

Kangaroos, 233.
Kerner, 146.
Korte, Dr. 117.
Kowalewsky, 199, 219, 251.

Labyrinthodonta, 72.
Lacertilia, 262.
Lama, 223, 245.
Lamarck, 85, 124, 148.
Lamellibranchiata, 199.
Lancelet, 36, 150, 219, 251.
Laplace, 15.
Leibnitz, 2.
Lemurs, 269, 280.
Lepidosirens, 37, 238, 258.
Leptalidæ, 180.
Leptotherium, 245.
Linnæus, 5, 84.
Lions, 223.
Lizards, 185, 261.
Locke, 303.
Lories, 233.
Luca, 289.
Luthardt, 12.
Lyell, Sir C. 123.

Machairodus, 242.
Macrauchenidæ, 273.
Madrepores, 42.
Mammals, 73, 240, 250, 264, 269, 277.
Mammoths, 79, 244, 247.
Man, 111, 201, 269, 288.
Mantidæ, 181.
Marsh, 267.
Marsipobranchii, 256.
Marsupial frog, 214.
Marsupials, 73, 75, 228, 238, 250, 269.

Martens, 241, 247.
Mastodon, 81.
Maupertuis, 4.
Maury, 227.
Mayer, Ernst, 118.
Medusæ, 31, 202, 210, 218.
Megalonyx Jeffersoni, 246.
Mollusca, 33, 199.
Monera, 27.
Monkeys, 79, 234, 269.—*See* Apes.
Monotremata, 269.
Müller, Friedrich, 305.
Müller, H. 299.
Müller, Johannes, 279.
Müller, Max, 161, 208, 238, 303.
Musk animals, 242.
Mylodon Harlemi, 246.
Myriapoda, 32.
Myxine, 258.

Nageli, 160, 193.
Nathusius, H. von, 175, 178.
Naumayr, 97.
Nauplius, 207, 210.
Navicula, 199.
Nematoids, 206.

Oken, 105.
Ophiura, 198.
Opossums, 233, 245.
Orang, 288.
Orniscelidæ, 260.
Ornithorhyncus, 37, 237, 270.
Orthoptera, 171.
Ostrich, 237.
Ouistitis, 291.
Ovibos, 277.
Owen, R. 121, 193, 390.
Oxen, 247, 275.
Oysters, 201.

Pachyderms, 79, 240, 242.
Palæochœridæ, 244.
Palæoniscus, 71.
Palæotheridæ, 273.
Paludina, 200.
Pander, 219.
Pavians, 289.
Peneus, 208.
Petromyzon, 256.
Phasmidæ, 181.
Physeteridæ, 280.
Pigs, 175, 178, 242, 244, 275.
Pigeons, 133, 170, 177.
Pinnipedæ, 278.
Placoids, 71.
Planorbis multiformis, 96.
Platelmintha Suctoria, 205.
Platypus, 233.
Plesiosaurians, 74, 260.
Pleuronectidæ, 181.
Polecats, 241.
Polistes Gallica, 47.
Polypes, 28, 30, 174, 203, 218.
Polypterus, 258.
Primates, 308.
Proboscidæ, 273.
Protamœba, 26, 40.
Proterosaurus, 73.
Proteus, 259.
Protopterus, 238, 258.
Pseudopus, 185.
Pterodactyls, 75, 250, 263, 265.
Pterotrachia, 199.
Pulmo-gasteropoda, 224.
Puma, 223.

Quadrumana, 245.

Radiata, 31.
Rathke, 55.
Regularæ, 77.
Reniera, 94, 154.
Reptiles, 250, 264.
Rhabdoliths, 25.

Rhinoceros, 231, 242, 247, 273, 310.
Rhizopoda, 207.
Rodents, 246, 269, 278, 280.
Rollet, 26, 168.
Rousseau, J. J. 4.
Ruminants, 79, 240, 277.
Rütimeyer, 81, 222, 227, 236, 257, 273, 277.

Sagitta, 37, 216, 219.
Sahuis, 289.
Saint Hilaire, E. G. 85.
Salamanders, 209, 211, 258.
Salmon, 239.
Sauria, 260.
Sauropsida, 264.
Scaphites, 76.
Schleicher, 304.
Schmidt, Johannes, 249.
Schulze, Max, 154.
Sea-cows, 277.
Sea-cucumbers, 77, 197.
Seals, 269.
Sea-snails, 198.
Sea-urchins, 76, 197.
Semper, 42.
Serpents, 260.
Sheep, 136, 244, 277.
Shrimp, 208.
Siebold, Von, 47.
Sirens, 238, 269, 277.
Sivatherium, 277.
Sloths, 228, 246.
Snails, 224.
Snakes, 185.
Spatangæ, 77, 80.
Spongiadæ, 30, 93, 218.
Squalodon, 279.
Starfish, 76, 197.
Stauridium, 42, 202.
Stein, 41.
Steinthal, 304.
Stone-lilies, 76, 80.
Strauss, D. F. 302.
Sturgeons, 238.
Suctoria, 206.
Suidæ, 276.
Surinam toad, 212.
Süssmilch, 4.

Tapeworm, 43, 205.
Tapirs, 231, 242, 273.
Tedania, 94.
Teleostei, 256.
Tellina, 96.
Termites, 178.
Tessellæ, 77.
Testacea, 250, 252.
Tetrabranchiata, 199.
Thompson, W. 64.
Threadworms, 206.
Thrushes, 234.
Tiger, 241.
Tœnia solium, 44.
Tortoises, 75, 238, 260, 263.
Tragulidæ, 275.
Trematoda, 205.
Trilobites, 69.
Tritons, 209, 259.
Trogons, 234.
Tuco-tuco, 184.
Turbellaria, 37, 45, 205.
Turrilites, 76.

Unger, F. 72.
Ungulates, 240, 242, 269, 273, 277.
Ursidæ, 278.

Verany, 182.
Vermes, 32, 201.
Vertebrata, 33, 154, 185, 250, 264.
Viverridæ, 241, 278.

Wagner, M. 158, 302.
Waitz, 300.
Wallace, 164, 230.
Walruses, 279.
Watson, H. C. 151.
Werner, 129.
Whales, 269, 280.
Wolves, 100.
Wombats, 233.
Woodpeckers, 234.
Worms, 205.
Würtenberger, 97, 213.

Zeuglodon, 279.
Zoea, 208.
Zöllner, 21, 162.

International Scientific Series.

D. Appleton & Co. have the pleasure of announcing that they have made arrangements for publishing, and have recently commenced the issue of, a Series of Popular Monographs, or small works, under the above title, which will embody the results of recent inquiry in the most interesting departments of advancing science.

The character and scope of this series will be best indicated by a reference to the names and subjects included in the subjoined list, from which it will be seen that the coöperation of the most distinguished professors in England, Germany, France, and the United States, has been secured, and negotiations are pending for contributions from other eminent scientific writers.

The works will be issued in New York, London, Paris, Leipsic, Milan, and St. Petersburg.

The International Scientific Series is entirely an American project, and was originated and organized by Dr. E. L. Youmans, who spent the greater part of a year in Europe, arranging with authors and publishers. The forthcoming volumes are as follows:

Prof. Lommel (University of Erlangen), *Optics.* (In press.)
Rev. M. J. Berkeley, M. A., F. L. S., and M. Cooke, M. A., LL. D., *Fungi; their Nature, Influences, and Uses.* (In press.)
Prof. W. Kingdon Clifford, M. A., *The First Principles of the Exact Sciences explained to the non-mathematical.*
Prof. T. H. Huxley, LL. D., F. R. S., *Bodily Motion and Consciousness.*
Dr. W. B. Carpenter, LL. D., F. R. S., *The Physical Geography of the Sea.*
Prof. William Odlong, F. R. S., *The Old Chemistry viewed from the New Standpoint.*
W. Lauder Lindsay, M. D., F. R. S. E., *Mind in the Lower Animals.*
Sir John Lubbock, Bart., F. R. S., *The Antiquity of Man.*
Prof. W. T. Thiselton Dyer, B. A., B. Sc., *Form and Habit in Flowering Plants.*
Mr. J. N. Lockyer, F. R. S., *Spectrum Analysis.*
Prof. Michael Foster, M. D., *Protoplasm and the Cell Theory.*
Prof. W. Stanley Jevons, *Money: and the Mechanism of Exchange.*
H. Charlton Bastian, M. D., F. R. S., *The Brain as an Organ of Mind.*
Prof. A. C. Ramsay, LL. D., F. R. S., *Earth Sculpture: Hills, Valleys, Mountains, Plains, Rivers, Lakes; how they were produced, and how they have been destroyed.*
Prof. Rudolph Virchow (Berlin University), *Morbid Physiological Action.*
Prof. Claude Bernard, *Physical and Metaphysical Phenomena of Life.*
Prof. H. Sainte-Claire Deville, *An Introduction to General Chemistry.*
Prof. Wurtz, *Atoms and the Atomic Theory.*
Prof. De Quatrefages, *The Negro Races.*
Prof. Lacaze-Duthiers, *Zoology since Cuvier.*
Prof. Berthelot, *Chemical Synthesis.*
Prof. J. Rosenthal, *General Physiology of Muscles and Nerves.*
Prof. James D. Dana, M. A., LL. D., *On Cephalization; or, Head-Characters in the Gradation and Progress of Life.*
Prof. S. W. Johnson, M. A., *On the Nutrition of Plants.*
Prof. Austin Flint, Jr., M. D., *The Nervous System and its Relation to the Bodily Functions.*
Prof. W. D. Whitney, *Modern Linguistic Science.*
Prof. C. A. Young, Ph. D. (of Dartmouth College), *The Sun.*
Prof. Bernstein (University of Halle), *Physiology of the Senses.*
Prof. Ferdinand Cohn (Breslau University), *Thallophytes (Algæ, Lichens, Fungi).*
Prof. Hermann (University of Zurich), *Respiration.*
Prof. Leuckart (University of Leipsic), *Outlines of Animal Organization.*
Prof. Liebreich (University of Berlin), *Outlines of Toxicology.*
Prof. Kundt (University of Strasburg), *On Sound.*
Prof. Rees (University of Erlangen), *On Parasitic Plants.*
Prof. Steinthal (University of Berlin), *Outlines of the Science of Language.*
E. Alglave (Professor of Constitutional and Administrative Law at Douai, and of Political Economy at Lille), *The Primitive Elements of Political Constitutions.*
P. Lorain (Professor of Medicine, Paris), *Modern Epidemics.*
Prof. Schützenberger (Director of the Chemical Laboratory at the Sorbonne), *On Fermentations.*
Mons. Debray, *Precious Metals.*

III.

Foods.

By Dr. EDWARD SMITH.

1 vol., 12mo. Cloth. Illustrated. . . . . Price, $1.75.

In making up THE INTERNATIONAL SCIENTIFIC SERIES, Dr. Edward Smith was selected as the ablest man in England to treat the important subject of Foods. His services were secured for the undertaking, and the little treatise he has produced shows that the choice of a writer on this subject was most fortunate, as the book is unquestionably the clearest and best-digested compend of the Science of Foods that has appeared in our language.

"The book contains a series of diagrams, displaying the effects of sleep and meals on pulsation and respiration, and of various kinds of food on respiration, which, as the results of Dr. Smith's own experiments, possess a very high value. We have not far to go in this work for occasions of favorable criticism; they occur throughout, but are perhaps most apparent in those parts of the subject with which Dr. Smith's name is especially linked."—*London Examiner.*

"The union of scientific and popular treatment in the composition of this work will afford an attraction to many readers who would have been indifferent to purely theoretical details. . . . Still his work abounds in information, much of which is of great value, and a part of which could not easily be obtained from other sources. Its interest is decidedly enhanced for students who demand both clearness and exactness of statement, by the profusion of well-executed woodcuts, diagrams, and tables, which accompany the volume. . . . The suggestions of the author on the use of tea and coffee, and of the various forms of alcohol, although perhaps not strictly of a novel character, are highly instructive, and form an interesting portion of the volume."—*N. Y. Tribune.*

IV.

Body and Mind.

THE THEORIES OF THEIR RELATION.

By ALEXANDER BAIN, LL. D.

1 vol., 12mo. Cloth. . . . . Price, $1.50.

PROFESSOR BAIN is the author of two well-known standard works upon the Science of Mind—"The Senses and the Intellect," and "The Emotions and the Will." He is one of the highest living authorities in the school which holds that there can be no sound or valid psychology unless the mind and the body are studied, as they exist, together.

"It contains a forcible statement of the connection between mind and body, studying their subtile interworkings by the light of the most recent physiological investigations. The summary in Chapter V., of the investigations of Dr. Lionel Beale of the embodiment of the intellectual functions in the cerebral system, will be found the freshest and most interesting part of his book. Prof. Bain's own theory of the connection between the mental and the bodily part in man is stated by himself to be as follows: There is 'one substance, with two sets of properties, two sides, the physical and the mental—a *double-faced unity.*' While, in the strongest manner, asserting the union of mind with brain, he yet denies 'the association of union *in place,*' but asserts the union of close succession in time,' holding that 'the same being is, by alternate fits, under extended and under unextended consciousness.'"—*Christian Register.*

D. APPLETON & CO., Publishers, 549 & 551 Broadway, N. Y.

V.

The Study of Sociology.

By HERBERT SPENCER.

1 vol., 12mo. Cloth. . . . . Price, $1.50.

"The philosopher whose distinguished name gives weight and influence to this volume, has given in its pages some of the finest specimens of reasoning in all its forms and departments. There is a fascination in his array of facts, incidents, and opinions, which draws on the reader to ascertain his conclusions. The coolness and calmness of his treatment of acknowledged difficulties and grave objections to his theories win for him a close attention and sustained effort, on the part of the reader, to comprehend, follow, grasp, and appropriate his principles. This book, independently of its bearing upon sociology, is valuable as lucidly showing what those essential characteristics are which entitle any arrangement and connection of facts and deductions to be called a science."—*Episcopalian.*

"This work compels admiration by the evidence which it gives of immense research, study, and observation, and is, withal, written in a popular and very pleasing style. It is a fascinating work, as well as one of deep practical thought."—*Bost. Post.*

"Herbert Spencer is unquestionably the foremost living thinker in the psychological and sociological fields, and this volume is an important contribution to the science of which it treats. . . . It will prove more popular than any of its author's other creations, for it is more plainly addressed to the people and has a more practical and less speculative cast. It will require thought, but it is well worth thinking about."—*Albany Evening Journal.*

VI.

The New Chemistry.

By JOSIAH P. COOKE, Jr.,

Erving Professor of Chemistry and Mineralogy in Harvard University.

1 vol., 12mo. Cloth. . . . . Price, $2.00.

"The book of Prof. Cooke is a model of the modern popular science work. It has just the due proportion of fact, philosophy, and true romance, to make it a fascinating companion, either for the voyage or the study."—*Daily Graphic.*

"This admirable monograph, by the distinguished Erving Professor of Chemistry in Harvard University, is the first American contribution to 'The International Scientific Series,' and a more attractive piece of work in the way of popular exposition upon a difficult subject has not appeared in a long time. It not only well sustains the character of the volumes with which it is associated, but its reproduction in European countries will be an honor to American science."—*New York Tribune.*

"All the chemists in the country will enjoy its perusal, and many will seize upon it as a thing longed for. For, to those advanced students who have kept well abreast of the chemical tide, it offers a calm philosophy. To those others, youngest of the class, who have emerged from the schools since new methods have prevailed, it presents a generalization, drawing to its use all the data, the relations of which the newly-fledged fact-seeker may but dimly perceive without its aid. . . . To the old chemists, Prof. Cooke's treatise is like a message from beyond the mountain. They have heard of changes in the science; the clash of the battle of old and new theories has stirred them from afar. The tidings, too, had come that the old had given way; and little more than this they knew. . . . Prof. Cooke's 'New Chemistry' must do wide service in bringing to close sight the little known and the longed for. . . . As a philosophy it is elementary, but, as a book of science, ordinary readers will find it sufficiently advanced."—*Utica Morning Herald.*

D. APPLETON & CO., Publishers, 549 & 551 Broadway, N. Y.

VII.

The Conservation of Energy.

By BALFOUR STEWART, LL. D., F. R. S.

With an Appendix treating of the Vital and Mental Applications of the Doctrine.

1 vol., 12mo. Cloth. Price, $1.50.

"The author has succeeded in presenting the facts in a clear and satisfactory manner, using simple language and copious illustration in the presentation of facts and principles, confining himself, however, to the physical aspect of the subject. In the Appendix the operation of the principles in the spheres of life and mind is supplied by the essays of Professors Le Conte and Bain."—*Ohio Farmer.*

"Prof. Stewart is one of the best known teachers in Owens College in Manchester.

"The volume of THE INTERNATIONAL SCIENTIFIC SERIES now before us is an excellent illustration of the true method of teaching, and will well compare with Prof. Tyndall's charming little book in the same series on 'Forms of Water," with illustrations encugh to make clear, but not to conceal his thoughts, in a style simple and brief."—*Christian Register, Boston.*

"The writer has wonderful ability to compress much information into a few words. It is a rich treat to read such a book as this, when there is so much beauty and force combined with such simplicity.—*Eastern Press.*

VIII.

Animal Locomotion;

Or, WALKING, SWIMMING, AND FLYING.

With a Dissertation on Aëronautics.

By J. BELL PETTIGREW, M. D., F. R. S., F. R. S. E., F. R. C. P. E.

1 vol., 12mo. . . . . . . Price, $1.75.

"This work is more than a contribution to the stock of entertaining knowledge, though, if it only pleased, that would be sufficient excuse for its publication. But Dr. Pettigrew has given his time to these investigations with the ultimate purpose of solving the difficult problem of Aëronautics. To this he devotes the last fifty pages of his book. Dr. Pettigrew is confident that man will yet conquer the domain of the air."—*N. Y. Journal of Commerce.*

"Most persons claim to know how to walk, but few could explain the mechanical principles involved in this most ordinary transaction, and will be surprised that the movements of bipeds and quadrupeds, the darting and rushing motion of fish, and the erratic flight of the denizens of the air, are not only anologous, but can be reduced to similar formula. The work is profusely illustrated, and, without reference to the theory it is designed to expound, will be regarded as a valuable addition to natural history."—*Omaha Republic.*

D. APPLETON & CO., PUBLISHERS, 549 & 551 Broadway, N. Y.

IX.

Responsibility in Mental Disease.

By HENRY MAUDSLEY, M. D.,

Fellow of the Royal College of Physicians; Professor of Medical Jurisprudence in University College, London.

1 vol., 12mo. Cloth. . . Price, $1.50.

"Having lectured in a medical college on Mental Disease, this book has been a feast to us. It handles a great subject in a masterly manner, and, in our judgment, the positions taken by the author are correct and well sustained."—*Pastor and People.*

"The author is at home in his subject, and presents his views in an almost singularly clear and satisfactory manner. . . . The volume is a valuable contribution to one of the most difficult, and at the same time one of the most important subjects of investigation at the present day."—*N. Y. Observer.*

"It is a work profound and searching, and abounds in wisdom."—*Pittsburg Commercial.*

"Handles the important topic with masterly power, and its suggestions are practical and of great value."—*Providence Press.*

X.

The Science of Law.

By SHELDON AMOS, M. A.,

Professor of Jurisprudence in University College, London; author of "A Systematic View of the Science of Jurisprudence," "An English Code, its Difficulties and the Modes of overcoming them," etc., etc.

1 vol., 12mo. Cloth. . . . . Price, $1.75.

"The valuable series of 'International Scientific' works, prepared by eminent specialists, with the intention of popularizing information in their several branches of knowledge, has received a good accession in this compact and thoughtful volume. It is a difficult task to give the outlines of a complete theory of law in a portable volume, which he who runs may read, and probably Professor Amos himself would be the last to claim that he has perfectly succeeded in doing this. But he has certainly done much to clear the science of law from the technical obscurities which darken it to minds which have had no legal training, and to make clear to his 'lay' readers in how true and high a sense it can assert its right to be considered a science, and not a mere practice."—*The Christian Register.*

"The works of Bentham and Austin are abstruse and philosophical, and Maine's require hard study and a certain amount of special training. The writers also pursue different lines of investigation, and can only be regarded as comprehensive in the departments they confined themselves to. It was left to Amos to gather up the result and present the science in its fullness. The unquestionable merits of this, his last book, are, that it contains a complete treatment of a subject which has hitherto been handled by specialists, and it opens up that subject to every inquiring mind. . . . To do justice to 'The Science of Law' would require a longer review than we have space for. We have read no more interesting and instructive book for some time. Its themes concern every one who renders obedience to laws, and who would have those laws the best possible. The tide of legal reform which set in fifty years ago has to sweep yet higher if the flaws in our jurisprudence are to be removed. The process of change cannot be better guided than by a well-informed public mind, and Prof. Amos has done great service in materially helping to promote this end."—*Buffalo Courier.*

XI.

Animal Mechanism,

A Treatise on Terrestrial and Aërial Locomotion.

By E. J. MAREY,

Professor at the College of France, and Member of the Academy of Medicine.

With 117 Illustrations, drawn and engraved under the direction of the author.

1 vol., 12mo. Cloth. . . . . Price, $1.75

"We hope that, in the short glance which we have taken of some of the most important points discussed in the work before us, we have succeeded in interesting our readers sufficiently in its contents to make them curious to learn more of its subject-matter. We cordially recommend it to their attention.

"The author of the present work, it is well known, stands at the head of those physiologists who have investigated the mechanism of animal dynamics—indeed, we may almost say that he has made the subject his own. By the originality of his conceptions, the ingenuity of his constructions, the skill of his analysis, and the perseverance of his investigations, he has surpassed all others in the power of unveiling the complex and intricate movements of animated beings."—*Popular Science Monthly.*

XII.

History of the Conflict between Religion and Science.

By JOHN WILLIAM DRAPER, M. D., LL. D.,

Author of "The Intellectual Development of Europe."

1 vol., 12mo. . . . . Price, $1.75.

"This little 'History' would have been a valuable contribution to literature at any time, and is, in fact, an admirable text-book upon a subject that is at present engrossing the attention of a large number of the most serious-minded people, and it is no small compliment to the sagacity of its distinguished author that he has so well gauged the requirements of the times, and so adequately met them by the preparation of this volume. It remains to be added that, while the writer has flinched from no responsibility in his statements, and has written with entire fidelity to the demands of truth and justice, there is not a word in his book that can give offense to candid and fair-minded readers."—*N. Y. Evening Post.*

"The key-note to this volume is found in the antagonism between the progressive tendencies of the human mind and the pretensions of ecclesiastical authority, as developed in the history of modern science. No previous writer has treated the subject from this point of view, and the present monograph will be found to possess no less originality of conception than vigor of reasoning and wealth of erudition. . . . The method of Dr. Draper, in his treatment of the various questions that come up for discussion, is marked by singular impartiality as well as consummate ability. Throughout his work he maintains the position of an historian, not of an advocate. His tone is tranquil and serene, as becomes the search after truth, with no trace of the impassioned ardor of controversy. He endeavors so far to identify himself with the contending parties as to gain a clear comprehension of their motives, but, at the same time, he submits their actions to the tests of a cool and impartial examination."—*N. Y. Tribune.*

DESCRIPTIVE SOCIOLOGY.

Mr. Herbert Spencer has been for several years engaged, with the aid of three educated gentlemen in his employ, in collecting and organizing the facts concerning all orders of human societies, which must constitute the data of a true Social Science. He tabulates these facts so as conveniently to admit of extensive comparison, and gives the authorities separately. He divides the races of mankind into three great groups: the savage races, the existing civilizations, and the extinct civilizations, and to each he devotes a series of works. The first installment,

THE SOCIOLOGICAL HISTORY OF ENGLAND,

in seven continuous tables, folio, with seventy pages of verifying text, is now ready. This work will be a perfect Cyclopædia of the facts of Social Science, independent of all theories, and will be invaluable to all interested in social problems. Price, five dollars. This great work is spoken of as follows:

From the British Quarterly Review.

"No words are needed to indicate the immense labor here bestowed, or the great sociological benefit which such a mass of tabulated matter done under such competent direction will confer. The work will constitute an epoch in the science of comparative sociology."

From the Saturday Review.

"The plan of the 'Descriptive Sociology' is new, and the task is one eminently fitted to be dealt with by Mr. Herbert Spencer's faculty of scientific organizing. His object is to examine the natural laws which govern the development of societies, as he has examined in former parts of his system those which govern the development of individual life. Now, it is obvious that the development of societies can be studied only in their history, and that general conclusions which shall hold good beyond the limits of particular societies cannot be safely drawn except from a very wide range of facts. Mr. Spencer has therefore conceived the plan of making a preliminary collection, or perhaps we should rather say abstract, of materials which when complete will be a classified epitome of universal history."

From the London Examiner.

"Of the treatment, in the main, we cannot speak too highly; and we must accept it as a wonderfully successful first attempt to furnish the student of social science with data standing toward his conclusions in a relation like that in which accounts of the structures and functions of different types of animals stand to the conclusions of the biologist."

www.ingramcontent.com/pod-product-compliance
Lightning Source LLC
LaVergne TN
LVHW010150110826
845151LV00002B/476

* 9 7 8 1 4 2 5 5 3 7 5 3 1 *